5th International Conference on Nanomaterials and Materials Engineering (ICNME 2017)

AA006701

IOP Conference Series: Materials Science and Engineering
Volume 204

Bali, Indonesia
1 – 3 April 2017

ISBN: 978-1-5108-4095-9
ISSN: 1757-8981

Printed from e-media with permission by:

Curran Associates, Inc.
57 Morehouse Lane
Red Hook, NY 12571

Some format issues inherent in the e-media version may also appear in this print version.

Copyright© (2017) by the Institute of Physics
All rights reserved. The material featured in this book is subject to IOP copyright protection, unless otherwise indicated.

Printed by Curran Associates, Inc. (2017)

For permission requests, please contact the Institute of Physics at the address below.

Institute of Physics
Dirac House, Temple Back
Bristol BS1 6BE UK

Phone: 44 1 17 929 7481
Fax: 44 1 17 920 0979

techtracking@iop.org

Additional copies of this publication are available from:

Curran Associates, Inc.
57 Morehouse Lane
Red Hook, NY 12571 USA
Phone: 845-758-0400
Fax: 845-758-2633
Email: curran@proceedings.com
Web: www.proceedings.com

Table of contents

Volume 204

5th International Conference on Nanomaterials and Materials Engineering

1–3 April 2017, Bali, Indonesia

Accepted papers received: 18 May 2017
Published online: 25 May 2017

Preface

Structure and Performance Analysis of Electrical Materials

Advanced Technology of Composites and Nanomaterials

Synthesis and Preparation of Materials: Modeling and Experiments

IOP Conf. Series: Materials Science and Engineering **204** (2017) 011001 doi:10.1088/1757-899X/204/1/011001

Preface

It is our great pleasure to welcome you to 2017 5th International Conference on Nanomaterials and Materials Engineering (ICNME 2017) which will be held in Bali, Indonesia during April 1-3, 2017. ICNME2017 is dedicated to issues related to Nanomaterials and Materials Engineering topics.

The major goal and feature of the conference is to bring academic scientists, engineers, industry researchers together to exchange and share their experiences and research results, and discuss the practical challenges encountered and the solutions adopted. Professors from different areas are invited to deliver keynote speeches regarding latest information in their respective expertise areas. It will be a golden opportunity for the students, researchers and engineers to interact with the experts and specialists to get their advice or consultation.

This proceedings present a selection from papers submitted to the conference from universities, research institutes and industries. The papers selected depended on their quality and their relevancy to the conference. The volume tends to present to the readers the recent advances in the field of Nanomaterials and Materials Engineering and various related areas, such as Biomaterials and Biomedical Engineering, Crystal Defect Engineering, Electronic and Magnetic Materials, Magnetic Materials, Materials Chemistry, Shape Memory Alloys and Martensites, etc..

We would like to thank all the authors who have contributed to this volume and also to reviewers, speakers, chairpersons, sponsors and all the conference participants for their support to ICNME 2017.

ICNME 2017 Organizing Committee

April 1-3, 2017

Content from this work may be used under the terms of the Creative Commons Attribution 3.0 licence. Any further distribution of this work must maintain attribution to the author(s) and the title of the work, journal citation and DOI.
Published under licence by IOP Publishing Ltd

IOP Conf. Series: Materials Science and Engineering **204** (2017) 011002 doi:10.1088/1757-899X/204/1/011002

Committee Members

Conference Co-Chairs

Dr. Tjokorda Gde Tirta Nindhia, Udayana University, Jimbaran, Bali, Indonesia

Dr. Brian Yuliarto, Institut Teknologi Bandung, Indonesia

Dr. Ahmed Aly Diaa Mohammed Sarhan, University of Malaya, Malaysia

Dr. Ildoo Chung, Pusan National University, Korea

Conference Program Chairs

Dr. Seong Soo Choi, SunMoon University and Sungkyunkwan University, Korea

Dr. Vladimir Tsepelev, Ural Federal University, Russia

Dr. Ahmed A D Sarhan, University of Malaya, Malaysia

Conference Technical Committees

Dr. Siti Rahmah Aid, Universiti Teknologi Malaysia, Malaysia

Dr.Hassan Soleimani, Universiti Teknologi PETRONAS (UTP), Malaysia

Dr.Nandini Sharma, University of Delhi, India

Dr.Deepak Kumar, IIT Delhi, India

Dr.Jonie Tanijaya, Paulus Christian University of Indonesia, Indonesia

Dr. Arnold Lubguban, Mindanao State University-Iligan Institute of Technology (MSU-IIT), Philippines

Dr. Aidy Ali, National Defence University of Malaysia, Malaysia

Dr. Gunawarman, Universitas Andalas, Indonesia

Dr. Jeyaraman Palanichamy, Kamaraj College of Engineering and Technology, India

Dr. Maziyar Sabet, Universiti Teknologi Brunei (UTB), Brunei Darussalam

Dr. Mohd Ani, International Islamic University of Malaysia, Malaysia

Dr. Munir Tasdemir, Marmara University, Turkey

Dr. Mohd Ani, International Islamic University of Malaysia, Malaysia

Dr. T.Theivasanthi, Kalasalingam University, India

Dr. S.Balasivanandha Prabu, Anna University, India

Dr. Anita S Ethiraj, VIT University, Vellore, India

Content from this work may be used under the terms of the Creative Commons Attribution 3.0 licence. Any further distribution of this work must maintain attribution to the author(s) and the title of the work, journal citation and DOI.
Published under licence by IOP Publishing Ltd

IOP Conf. Series: Materials Science and Engineering **204** (2017) 011002 doi:10.1088/1757-899X/204/1/011002

Dr. Abbas Teimouri, Payame Noor University (PNU), Iran
Dr. Osman Adiguzel, Firat University, Elazig / Turkey
Dr. H.Özkan GÜLSOY, Marmara University, Turkey
Dr. P.Sivaprakash, A.S.L.Pauls College of Engineering &Technology, India
Dr. Geetesh Goga, K.C. College of Engg. & I.T., Nawanshahr, Punjab, India
Dr. Angshuman Khan, University of Engineering & Management, India
Dr. Abdul Razak Ismail, Universiti Teknologi Malaysia, Malaysia
Dr. Madya Issham Ismail, Universiti Teknologi Malaysia, Malaysia
Dr. Azmah Hanim Mohamed Ariff, Universiti Putra Malaysia
Dr. Nirali Gondaliya, SVM institute of technology, gujarat, India
Dr. Arjon Turnip, Indonesian Institute of Sciences, Indonesia
Dr. Hendra Suherman, Universitas Bung Hatta, West Sumatera, Indonesia

IOP Conf. Series: Materials Science and Engineering **204** (2017) 011003 doi:10.1088/1757-899X/204/1/011003

Peer review statement

All papers published in this volume of *IOP Conference Series: Materials Science and Engineering* have been peer reviewed through processes administered by the proceedings Editors. Reviews were conducted by expert referees to the professional and scientific standards expected of a proceedings journal published by IOP Publishing.

Content from this work may be used under the terms of the Creative Commons Attribution 3.0 licence. Any further distribution of this work must maintain attribution to the author(s) and the title of the work, journal citation and DOI.
Published under licence by IOP Publishing Ltd

IOP Conf. Series: Materials Science and Engineering **204** (2017) 012001 doi:10.1088/1757-899X/204/1/012001

Structural and magnetic properties characterization of mechanically alloyed La substituted $Ba_{(1-x)}La_xFe_{12}O_{19}$ (x = 0.07, 0.15 and 0.20)

Kilat Permana Putra andAzwar Manaf

Postgraduate Program of Materials Science Study, Faculty of Mathematics and Natural Sciences, Universitas Indonesia, Depok 16424, Indonesia

Email :azwar@ui.ac.id

Abstract. This study reported the structural studies and magnetic properties of La-substituted barium hexaferrite (BHF) with $Ba_{1-x}La_xFe_{12}O_{19}$ (x=0. 07, 0.15 and 0.20) constitutions. All samples were synthesized through mechanical alloying technique using $BaCO_3$, La_2O_3 and Fe_2O_3precursors that were mechanically milled for 20 hours and successively sintered at 1200 ^{0}C to produce polycrystalline materials. The X-ray diffraction (XRD) method was employed to characterize the microstructure of synthesized materials. Structural analysis by refinement of XRD data showed that all sintered $Ba_{1-x}La_xFe_{12}O_{19}$ (x=0.07, 0.15 and 0.20) samples are single phase material.The crystallite size of all samples was below 150 nm, with the mean size value decreased from 35 nm in sample of x = 0.07 to 23 nm in x = 0.20. This was accompanied by a change in crystallite size distribution toward the slim distribution.Magnetic properties characterization of La substituted BHF showed a substantial increase of coercivity as the fraction of La increased but the remanent magnetization decreased.The largest coercivity value of 315 kA/m was obtained in sample with x = 0.20 which is almost double when compared with low substitution of La (x = 0.07).

1. Introduction

Hexagonal Barium or strontium ferrites, $(Ba/Sr)Fe_{12}O_{19}$ [1] are still attractive compounds for many engineers and researchers due to their low costs and outstanding chemical stability. They have superior magnetic properties like high values of saturation magnetization, Curie temperature, crystalline anisotropy constant and high corrosion resistance [2,3]. In spite of hexagonal ferrites have met many applications in wide range area like permanent magnets, radar absorbing materials, sensors and recording media [1], improvement in magnetic properties still be of great industrial expectation, which caused research activities in this type of compound still kept going.There have been some treatments, which are reported to increase the magnetic properties of hexaferrite through microstructure refinement toward nanocrystalline regime or substitution divalent or trivalent ions. The high value of coercivity may be obtained in materials with crystallite size closer to the single domain particle [4,5].The critical size of single domain particles for the barium hexaferrite is 460 nm [6]. However, when the crystallite size is significantly lower than single domain particles, the exchange interaction effect could decreases the coercivity due to a change in the effective anisotropy constant [7].

There are some methods, offering to produce ceramic materials barium hexaferrite like mechanical alloying [8], sol-gel [9], hydrothermal method [10], solution combustion [11]. In spite of successes the above synthesis methods, our current research is aimed at finding out the changes in microstrure and magnetic properties caused by a partial substitution of lanthanum to barium in $Ba_{1-x}La_xFe_{12}O_{19}$ (x=0,

Content from this work may be used under the terms of the Creative Commons Attribution 3.0 licence. Any further distribution of this work must maintain attribution to the author(s) and the title of the work, journal citation and DOI.
Published under licence by IOP Publishing Ltd

07–0.20) through mechanical alloying method. The substituted of La^{3+} ion is expected to increase the coercive field strength. Structural and magnetic properties of synthesized $Ba_{1-x}La_xFe_{12}O_{19}$ (x=0,07–0.20) were evaluated respectively by XRD studies and permagraph. The Rietveld method together with *whole powder pattern modelling* (WPPM) [12] were employed to study the microstructure.

2. Experimental Methods

Synthesis of La-substituted barium hexaferrite with $Ba_{1-x}La_xFe_{12}O_{19}$ (x=0,07-0,20) compositions was carried out under mechanical alloying method. $BaCO_3$, Fe_2O_3 and La_2O_3 precursors with each purity level better than 99 % were mixed and milled in a planetary ball mill for 20 hours under wet milling condition. The ball to powder mass ratio was 10:1. The mechanically milled powders were then loaded into a cylindrical die of 25 mm in diameter and successively pressed with 10 tons uniaxial force to produce a green compact in form of pellet tablet. The green compacts of each sample were sintered at a temperature of 1200 ^{0}C for 2 hours to obtain crystalline samples.Phases in sintered samples were then identified by XRD from which the XRD data were analyzed further using structural analysis software, GSAS.The crystallite size distribution of sintered samples of $Ba_{1-x}La_xFe_{12}O_{19}$ (x=0, 07-0,20) compositions was calculated employingWPPM. Their corresponding magnetic properties were evaluated by a permagraph.

3. Results And Discussion

Plot of XRD pattern of crystalline $Ba_{1-x}La_xFe_{12}O_{19}$ (x=0,07; 0,15; 0,20) samples is compared in Figure 1.Careful identification to the diffraction peaks in the diffraction angle range 10-80^{0} for all patterns confirmed the three samples have a similar pattern. but a small shift in peaks position was observed (see inset in Figure 1).

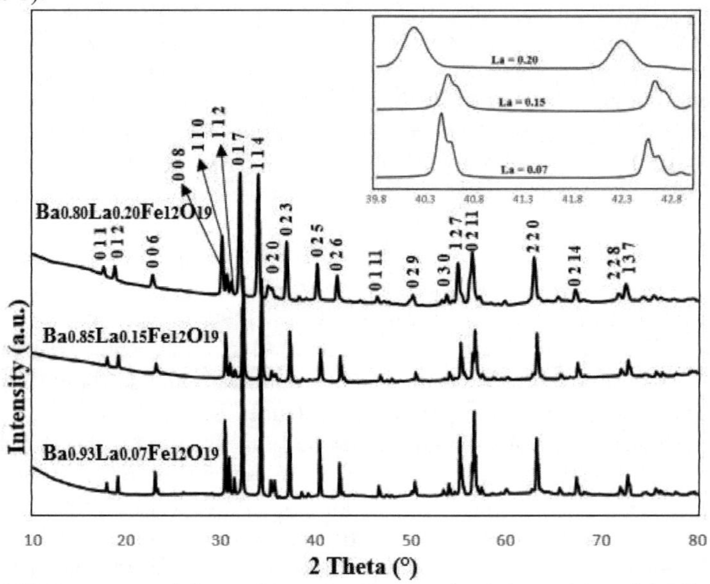

Figure 1.XRD pattern of sintered $Ba_{1-x}La_xFe_{12}O_{19}$ (x=0.07, 0.15 and 0.20) samples.

XRD data processing by structural analysis software for phase identification and crystal parameter determination, referred to the convergence fitting process after acceptable goodness of fit values are achieved. Results of XRD data refinement for $Ba_{1-x}La_xFe_{12}O_{19}$ (x=0.07-0.20) samples were summarizedin table 1. The formation of a single phase barium hexaferritewas indicated by very good fit between the model and experimental data which was verified by the values of fitting parameters like *Gof* and *R-factors*. The experimental data were compared with crystalography open database #1008841 for barium hexaferrite.This indicates that the La^{3+} ions were completely diluted in the hexagonal crystal lattice of BHF. A small shift in the peak positions to the higher diffraction angles occurredin a sample with $La^{+3}$is 0.15. However, the shift the peak positions change to the lower angle when the substitution of $La^{+3}$is 0,20. The peak shifting must be due to the presence of La^{+3} ion in partially substitute Ba^{+3} of BHF.Referring tothe listed parameters showed in the Table 1, there is a small change in the calculated lattice constants of La-substituted BHF samples. Obviously, this occurred due to the differences in ionic radii between La^{+3} and Ba^{+3} in BHF. Consequently, the volume of unit cell also changed slightly.

Table 1. Fitting parameters $Ba_{1-x}La_xFe_{12}O_{19}$.

X	*Gof*	*Rp*	*wRp*	a(Å)	c(Å)
0.07	1.05	2.92	3.65	5.8870	23.2188
0.15	1.11	3.00	3.77	5.8905	23.1926
0.20	1.05	2.46	3.11	5.8882	23.1696

The XRD data of diffraction patterns in Figure 1 were further utilizedfor microstructure analysis like crystallite size distribution and mean crystallite size determination. The crystallite size distribution which calculated by the *whole powder pattern modelling* software of samples with various La^{3+} content in La substituted BHF is plotted in Figure 2. The calculation was based on the assumption of a spherical crystallite shape, which is suitable for nanocrystalline materials. Distribution curves in Figure 2 indicated that the increase in ionic fraction of La has changed the crystallite size distribution toward a more slime distribution, in which the crystallites with a low mean value increased.La^{3+} substitution gives an effect of a decrease in the width of the crystallite size distribution and hence decreased the mean crystallite size. The latter was found 23 nm in sample with x = 0.20 which is almost half of the mean crystallite size of sample with x = 0.07. This may be associated with a faster crystallite growth rate in La substituted BHF samples.

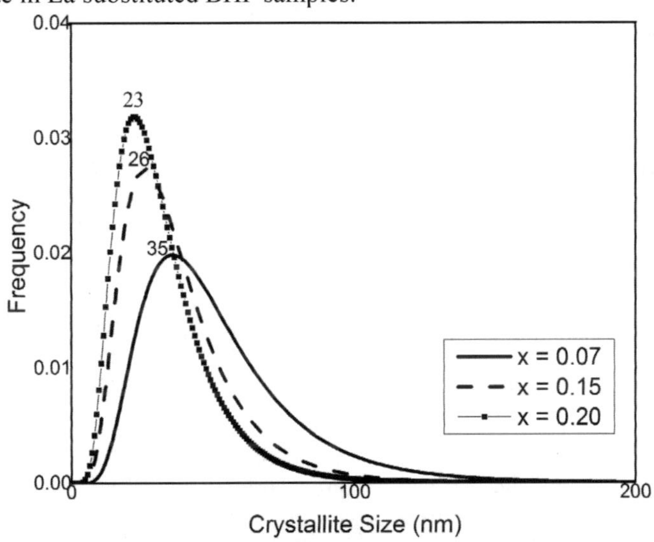

Figure 2. Distribution crystallite size dependence La content.

Hysteresis loops of lanthanum substituted BHF samples were plotted in Figure 3. All doped samples characterized by typical of permanent magnets. For sample with a small amount of La substitution in BHF (x = 0.07), the loop is almost identical with that of un-doped BHF (x = 0). The magnetization value that obtained at a maximum applied magnetic field, M(T)for this particular sample is 0.33 Twhich was then extrapolated by Law of Approach to Saturation method [14] for total magnetization (M_s) value. Remanence and coercivity values are respectively 0.19 T and 204.8 kA/m. These magnetic property values are typical of the isotropic BHF permanent magnet, though the M(T) value is a slighly lower than that of BHF (0.39 T). The loops also exhibited an increase in the coercivity with an increase in the amount of La substitution, being the largest coercivity obtained was in a sample with the La^{3+} content. For the sample with x = 0.20, the coercivty was 314.9 kA/m accompanied by a decrease in remanent magnetization value from 0.19 T to 0.14 T. Table 2 summarized magnetic properties of $Ba_{1-x}La_xFe_{12}O_{19}$ (x=0.07, 0.15 and 0.20) samples.

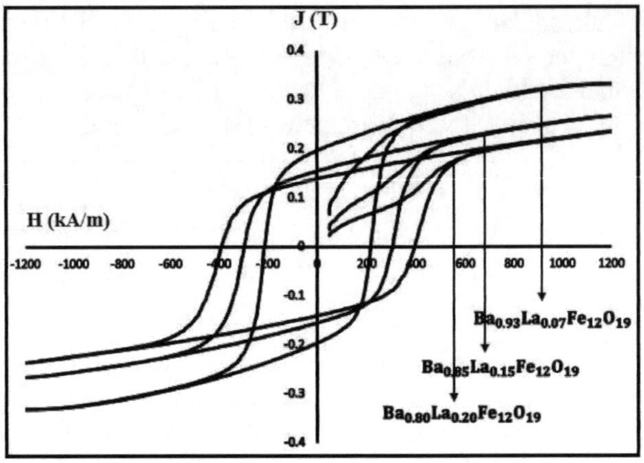

Figure 3. Hysteresis loopfor $Ba_{1-x}La_xFe_{12}O_{19}$($x = 0.07, 0.15, 0.20$) samples.

The results of the magnetic property evaluation as discussed above have shown that the substitution of La^{3+} in BHF affected the coercive field strength (H_c), remanence (M_r) and total magnetization (M_s). The magnetic properties of the whole samples derived from the hysteresis loopsas summarized in Table 2 have assured that samples with La substitution in BHF (x = 0.07-0.20) are inferior in M and Mr values, but superior in Hc values, when compared with un-doped BHF. The enhancement in coercivity value must be due to a change to a large value of magnetocrystalline anisotropy constant with La substitution. Obviously, a decrease in remanent must be due to a decrease in total magnetization value of La-substituted BHF samples. This was in agreement with the results reported by Ovalioglue et.al showed the hysteresis loops of the Ba1-xLaxFe12O19 hexaferrite with x=0.00-0.40, *Ms* decreased with increasing La substituted up to x=0.40 [13].

Table 2. Detail magnetic properties of lanthanum substituted barium hexaferrite.

X	M_s(T)	M_r(T)	M_r/M_s	H_c (kA/m)	$(BH)_{max}$ (kJ/m^3)
0.00	0.35	0.19	0.54	197.7	6.2 [15]
0.07	0.36	0.19	0.53	204.8	6.2
0.15	0.29	0.15	0.52	269.0	4.2
0.20	0.25	0.14	0.56	314.9	3.4

Included in Table 2, their respective maximum energy product, $(BH)_{max}$ value. The $(BH)_{max}$ of sample with x = 0.00 and 0.07 is similar, obviously because the two samples have the same remanence value. It is worth to note that despite samples have nanostructured with mean crystallite size in nanorezime(23-35 nm) and far below the single domain crystallite size of BHF (460 nm), no remanence enhancement was observed. In addition, the calculated exchange length in BHF was reported about 6 nm [16]. With such length, the grain exchange interaction should take place [17] and remanence enhancement could be expected rise in the all samples. However, this enhancement was not obtained from the samples.

4. Conclusion

The microstructure and magnetic properties of lanthanum substituted BHF samples with $Ba_{1-x}La_xFe_{12}O_{19}$ (x=0.07, 0.15 and 0.20) compositions were investigated. Microstructurally, all samples are single phase material with crystallite size in nanocrystallinerezime. It was shown that the increase in La^{+3} substitution is to squeeze the width of the crystallite size distribution toward a slim distribution with the mean crystallite size of 35 nm in x = 0.07 shifted to 23 nm in x = 0.20. The magnetic properties of the whole samples have assured that samples with La substitution in BHF are inferior in M and Mr values, but superior in Hc values, when compared with un-doped BHF. No remanence enhancement was observed.

5. Acknowledgements

IOP Conf. Series: Materials Science and Engineering **204** (2017) 012001 doi:10.1088/1757-899X/204/1/012001

The authors gratefully acknowledge the support of Postgraduate Program of Materials Science Universitas Indonesia for the research facilities. We are also thankful for the financial support provided by the Directorate of Research and Community Service Universitas Indonesia under program Grants of International Publication Indexed for Final Project Student 2016 (PITTA 2016) contract no.1994/UN2.R12/HKP.05.00/2016.

References

[1] R. Nowosielski, R. Babilas, G. Dercz and L. Pajak 2008 *journal of achievements in materials and manufacturing engineering* **31**, issue 2 pp 269-274.
[2] L. A. Dobrzanski, M. Drak and B. Ziebowicz 2006 *journal of achievements in materials and manufacturing engineering* **17** pp 37-40.
[3] R. M. Kershi and Samir Osman Al- Asbahi 2013*International Journal of Chemical, Environmental & Biological Sciences (IJCEBS)* **1**, Issue 1ISSN pp 2320–4087.
[4] H. Stablein and E.P. Wohlfarth 1982*Ferromagnetic Materials* **3**(North-Holland: Amsterdam).
[5] J. Ding, H. Yang, W.F. Miao, P.G. McCormick and R. Street 1995*J. Alloys Comp.* **221** p 70.
[6] Rezlescu, E. Rezlescu, P. D. Popa and N. Rezlescu 1997*J. Magn. Magn. Mater* **193** p. 288.
[7] Herzer G 1990*IEEE Trans. Magn,***26** pp 1397-1402.
[8] R. Nowosielski,R. Babilas, G. Dercz, L. Pajak and J. Wrona2007 *international scientific journal* **28**, issue 12 pp735-742.
[9] Li Jie, Zhang Huai-Wu, Li Yuan-Xun, Liu Ying-Li and Ma Yan-Bing 2012*Chin. Phys. B* **21**, No. 1 017501.
[10] Giap Vi. Duong, R. Sato turtelli, B. D. Thuan, D. V. Linh, N. Hanh and R. Groessinger 2007 *Journal of Non-Crystalline Solids* **353** pp 811–813.
[11] Vinod N. Dhage, M.L. Mane, A.P. Keche, C.T. Birajdar and K.M. Jadhav 2011*Physica* ***B*406** pp 789–793.
[12] Scardi, P and Leoni 2002*Whole powder pattern modelling research papers* **190–200**.
[13] H. Ovalioglu, H. Sozeri, M. Kabaer and I. Kucuk 2010 *ACTA PHYSICA POLONICA A* No. 5 **118** pp 1020-1021.
[14] Grössinger, R. 1982*Journal of Magnetism and Magnetic Materials* ***28*(1-2)** pp 137–142.
[15] Kilat Permana Putra and Azwar Manaf2016 *the 2nd international symposium on current progress in mathematics and sciences*(unpublished) AIP Publishing.
[16] J. M. D. Coey 2009*Magnetism and Magnetic Materials*, 234-249, 418-427, 472 (Cambridge University Press, New York).
[17] Manaf, R.A. Buckley and H.A. Davies 1993 *J. Magn. Magn. Mater.*, **128** pp 307-312.

IOP Conf. Series: Materials Science and Engineering **204** (2017) 012002 doi:10.1088/1757-899X/204/1/012002

Effect of pH on ionic liquid mediated synthesis of gold nanoparticle using elaiseguineensis (palm oil) kernel extract

Muhammad Irfan[1], Tausif Ahmad[1], Muhammad Moniruzzaman[1,2]and Bawadi Abdullah[1]

[1]Chemical Engineering Department, Universiti Teknologi PETRONAS, Bandar Seri Iskandar, 32610, Perak, Malaysia

E-mail: sahalirfan@gmail.com, tausif_engr@yahoo.com, m.moniruzzaman@utp.edu.my, bawadi_abdullah@utp.edu.my

Abstract. This study was conducted for microwave assisted synthesis of stable gold nanoparticles (AuNPs) by reduction of chloroauric acid with Elaeis Guineensis (palm oil) kernel (POK) extract which was prepared in aqueous solution of ionic liquid, [EMIM][OAc], 1-Ethyl-3-methylimidazolium acetate. Effect of initial pH of reaction mixture (3.5 - 8.5) was observed on SPR absorbance, maximum wavelength (λ_{max}) and size distribution of AuNPs. Change of pH of reaction mixture from acidic to basic region resulted in appearance of strong SPR absorption peaks and blue shifting of λ_{max} from 533 nm to 522 nm. TEM analysis revealed the formation of predominantly spherical AuNPs with mean diameter of 8.51 nm. Presence of reducing moieties such as flavonoids, phenolic and carboxylic groups in POK extract was confirmed by FTIR analysis. Colloidal solution of AuNPs was remained stable at room temperature and insignificant difference in zeta value was recorded within experimental tenure of 4 months.

1. Introduction

The synthesis of gold nanoparticles (AuNPs) has got immense attention due to their ingenious applications in various industrial sectors such as environment science & engineering, catalysis, optical fibers, food & pharma, electronics and medical science [1]. Conventionally, synthesis of AuNPs can be performed by many physical, chemical and green synthesis methods [2]. However, high cost and toxic chemicals are major challenges for applying physical and chemical methods therefore bio synthesis got edge due to its simple, eco-friendly and low cost process [3]. Phytochemicals present in biomass are efficient to reduce chloroauric acid to form AuNPs. In fact, synthesis of nanomaterials using bio-synthesis approach has provided a common junction between nanotechnology and bio technology which led to the fabrication of unique materials with tunable size and shapes.

The role of pH is significant in changing the size and shape of nanoparticles. Numbers of studies has shown the stability of AuNPs at basic pH while many achieved stable suspension in acidic region. Olive leaf extract was used to synthesize stable AuNPs at basic pH, however, particles were precipitated within 12 h in acidic region [4]. Contrariwise, AuNPs synthesized using banana peel extract were stable at pH value of 2-5 [5]. A broader range of pH i.e. 2-11 was taken to synthesize AuNPs using oil palm mill effluent and pH 3 was observed optimum to achieve definite shapes

Content from this work may be used under the terms of the Creative Commons Attribution 3.0 licence. Any further distribution of this work must maintain attribution to the author(s) and the title of the work, journal citation and DOI.
Published under licence by IOP Publishing Ltd

IOP Conf. Series: Materials Science and Engineering **204** (2017) 012002 doi:10.1088/1757-899X/204/1/012002

particles [6]. Furthermore, the synthesis reaction, size, shape and stability of AuNPs could be controlled by adjusting the initial pH value of reaction mixture.

To get plant extract, traditionally, water is considered a cheap and eco-friendly solvent; however, synthesized AuNPs are sometime not much stable at room temperature and form aggregates diminishing the colloidal stability [7, 8]. Ionic liquids (ILs) are therefore considered as a substituent due to "green" nature and strong solvating interaction with polar and non-polar compounds [9]. ILs are composed of ions and have been reported as an efficient solvent to extract various phytochemicals over many other traditional solvents such as water, alcohol, acetone and ether [10]. These have also been used as stabilizers in conventional treatment method for synthesis of AuNPs [11].

Palm oil trees are abundantly available in Indonesia and Malaysia and could be a potential source for bio-synthesis of AuNPs due to occupying substantial amount of phytochemicals particularly flavonoids and phenolic compounds [12]. In this study, we have therefore aimed to synthesize AuNPs using palm oil kernel (POK) extract, in which dual role of ILs (i) for extraction of phytochemicals from POK and (ii) stabilizers for AuNPs were investigated in a single step. Moreover, influence of pH on the rate of reaction and morphological characteristics of AuNPs was also studied.

2. Materials and Methods

Gold (III) chloride hydrate ($HAuCl_4.3H_2O$) and IL[EMIM][OAc], 1-ethyl-3-methylimidazolium acetate were procured from Sigma-Aldrich. The palm oil kernels were collected from Felcra Berhad Nasaruddin Oil Palm Plantation located in Bota, Perak, Malaysia.

2.1 Preparation of POK extract

Fresh POKs were collected and washed with plenty of distilled water to remove dust contaminants. These POKs were dried in an oven at 70 °C for 12 h followed by grinding by using IKA® grinder with 0.25 mm sieve. To get POK extract, 10 g of finely grinded kernels were added to 100 mL of [EMIM][OAc] 2%(w/v) and heated at 90 °C for 20 min. The mixture was filtered via gravitational filtration using Whatman No. 1 filter paper. The filtrate was collected in 100 mL glass vial and stored at 4 °C for further use. This extract was used within three days for reaction with chloroauric acid to avoid any loss of reduction efficiency for synthesis of AuNPs [13].

2.2 Synthesis of AuNPs

Typical reaction mixture is prepared by drop wise addition of 2 mL of POK extract to an aqueous mixture of 2 mL of 2.28 mM chloroauric acid added to 10 mL distilled water. The resulting mixture was heated in a microwave oven (Sharp – R 268R/S-M) with an output power capacity of 800 W and frequency of 2450 MHz. Effect of pH was determined by using different prefixed initial pH values of reaction mixture (3.5 – 9.5). HCl or NaOH, 0.1 M were used to adjust pH value.

2.3 Characterization of AuNPs

Perkin Elmer Lambda 25 UV-Visible (*UV-Vis*) spectrophotometer was used to record SPR absorbance and maximum peak wavelength (λ_{max}) at 1 nm resolution and a scan speed of 480 nm/min. The FTIR spectrum was observed using Ziess Supra 55 CP FTIR spectrophotometer. FTIR analysis was performed for POK extract before and after its reaction with chloroauric acid to identify the bio-compounds that could involve in reduction of gold ions into aurum particles. Zeiss Libra 200 TEM was used to observe the morphological characteristics of synthesized AuNPs. Analysis was performed by putting one drop of colloidal gold solution on the copper grid and allowed to dry prior to TEM imaging and size measurement. To predict the charge stability of colloidal gold, zeta potential value was measured by using "Malvern, Zetasizer Nano-ZSP".

3. Results and Discussions

Initial colour of POK extract was translucent white however, an addition of the chloroauric acid and distilled water was resulted in a clear solution. After heating the mixture in MW oven for 2 min, colour of the solution turned into light pink which became more intense with time. This change of colour from white to pinkish is primary indication for successful formation of AuNPs due to excitation of surface plasmon vibration in the AuNPs.

3.1 Effect of pH on biosynthesis of AuNPs

Reaction was performed at pH 3.5 - 9.5 to identify the effect of pH on formation of AuNPs. It was observed that absorbance of solution increased while changing the initial pH of the solution from acidic to basic region as shown in figure 1. With increase of pH, SPR band was also blue shifted toward low wavelength region (figure 1 inset). *UV-vis* results suggested that no reaction was occurred in acidic region at pH 3.5 which is in agreement with the previous studies [14, 15]. It was might be due to protonation of hydroxyl and carboxyl group presents in extracts [16].

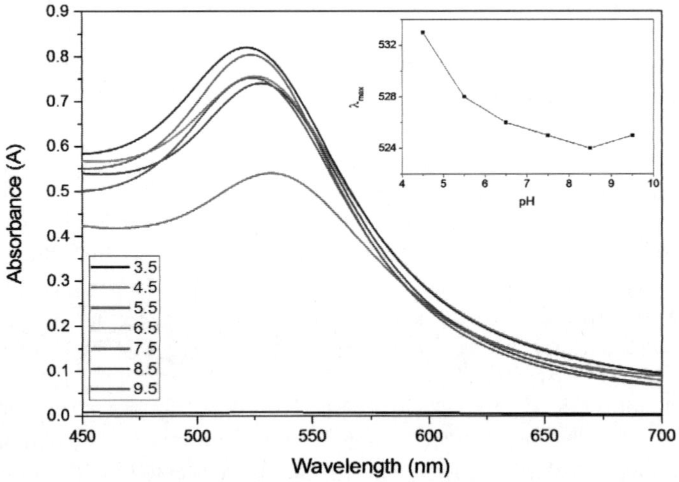

Figure 1: *UV-vis* spectra; effect of initial pH of reaction mixture (a) SPR peak positions (b) change of λ_{max}

Based on DLVO theory, first capping of AuNPs is done by anions therefore, anions are considered as a key source for enhanced colloidal stability [17]. However, strong acidic conditions protonated the acetate ions present in the reaction mixture and thus reducing their capping tendency and electrosteric forces to impart higher stability to particles which may resulted in slow reduction of gold ions or precipitation of AuNPs from solution. With increase of pH toward basic region, rapid change of colour was occurred with formation of sharp SPR peaks. Colloidal gold solution was remained stable at room temperature and there was nothing any aggregation or precipitation was noticed even after four months of their preparation indicating the strong stabilizing capability of IL.

TEM image analysis was performed for two representative sample of AuNPs synthesized at pH 4.5 and pH 7.5 as shown in figure 2a and 2b. The TEM images revealed that the AuNPs formed were predominantly spherical with narrow size distribution. At pH 4.5, the mean size of the AuNPs was 9.61 nm with size distribution ranged from 4.49 nm to 17.56 nm. However, when initial pH of reaction mixture was 7.5, mean sizes of the AuNPs obtained were relatively smaller. Particles were more uniform and well dispersed endorsing the validity of results driven from UV-vis. The size of the AuNPs ranged from 4.32 nm to 16.12 nm with the mean diameter of 8.51 nm as shown in figure 2b. Comparing the TEM images at both pH, it was obvious that AuNPs were more spherical, uniform and dispersed in basic region

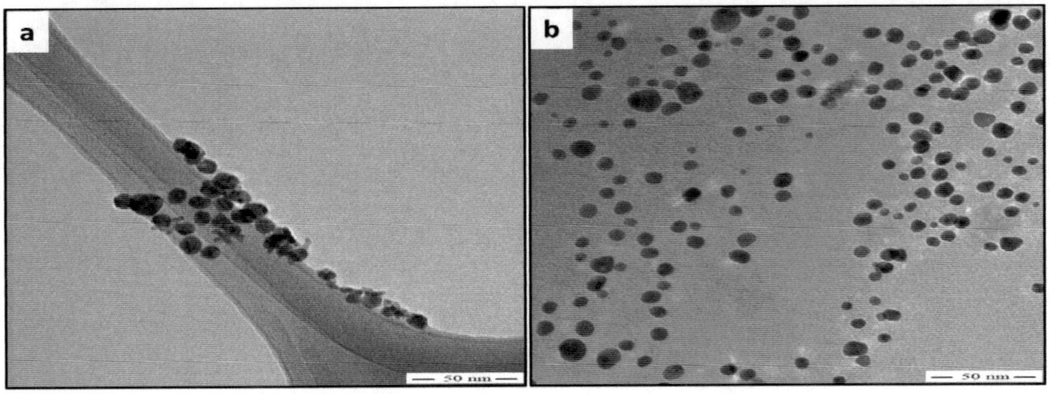

Figure 2: TEM image analysis of AuNPs at (a) pH 4.5, (b) 7.5

3.2 Zeta Potential

Colloidal stability of AuNPs depends upon the interaction between particles. A colloidal solution is stable when electrostatic repulsion forces are enough stronger than van der Waals attractive forces. Zeta potential value can be sued to identify the electrostatic repulsion. AuNPs are believed to be more stable if occupy zeta potential in between -30 mV to +30 mV [18]. The zeta value of synthesized AuNPs prepared at pH 7.5 was observed as -21.1 mV showing the good stability of colloidal solution. AuNPs solution was preserved at room temperature for four months and again tested for its zeta value which was found to be -19.5 mV as shown in figure 3. This value was still high and showed the enhanced colloidal stability of AuNPs.

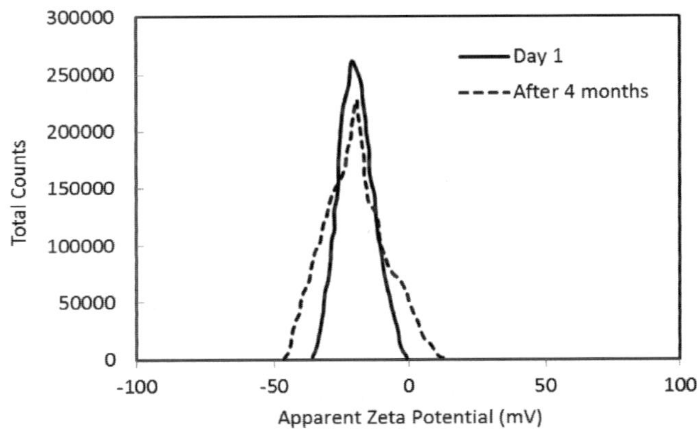

Figure 3: Zeta potential of AuNPs

3.3 FTIR Analysis

FTIR analysis was performed for POK extract and AuNPs mixture synthesized at pH 7.5 to identify the possible functional groups and other organic moieties involved in the synthesis of AuNPs. From the spectrum, various bands were observed at 691 cm^{-1}, 1114 cm^{-1}, 1642 cm^{-1} and 3430 cm^{-1} as shown in figure 4. These bands represented the presence of C-H aromatic, aliphatic amines, amide C=O and hydroxyl groups, respectively. A strong band at 3430 cm^{-1} indicated the presence of -OH groups which suggested the involvement of phenolic and flavonoids compounds as reducing agents for formation of AuNPs [19, 20]. Therefore, it could be presume that flavonoids and phenols which are abundant in palm oil biomass showed their characteristics peaks and seemed to be responsible for accelerated reduction of gold ions.

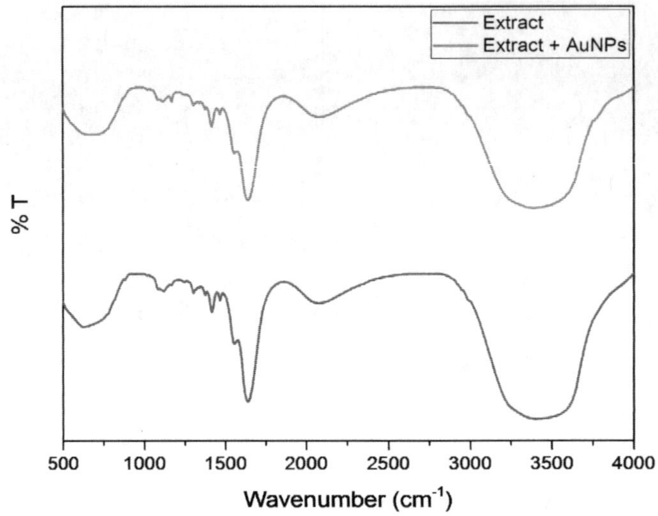

Figure 4: FTIR analysis for POK extract and AuNPs mixture

4. Conclusion

A single step method was established to synthesize AuNPs using POK extract prepared in [EMIM][OAc]. Substantial amount of flavonoids and phenolics compounds in POK were responsible to carry out the reduction reaction for bio-synthesis of AuNPs. Higher polarizability, strong MW absorbent and capping tendency of AuNPs made ILs as efficient solvents for rapid synthesis of stable AuNPs. The sizes and shape of synthesized AuNPs can also be tuned by manipulating pH of the reaction mixture. Colloidal gold solution was remained stable at room temperature without any aggregation. Higher stability of AuNPs was also exhibited by high zeta value of AuNPs even after four months.

Acknowledgement

The author would like to express his gratitude towards Universiti Teknologi Petronas for financial support through STIRF (0153AA-C77) and FRGS (0153AB-I96). Author would also graciously acknowledge the financial assistance through Graduate Assistance (GA) scheme by Universiti Teknologi Petronas, Malaysia for pursuing PhD studies.

References

[1] N. U. Islam, K. Jalil, M. Shahid, N. Muhammad, and A. Rauf, "Pistacia integerrima gall extract mediated green synthesis of gold nanoparticles and their biological activities," *Arab. J. Chem.*2015, http://dx.doi.org/10.1016/j.arabjc.2015.02.014.

[2] S. Ahmed and S. Ikram, "Biosynthesis of gold nanoparticles: A green approach," *J. Photochem. Photobiol.,B: Bio.***161**, pp. 141-53, 2016.

[3] A. Umer, S. Naveed, N. Ramzan, and M. S. Rafique, "Selection of a suitable method for the synthesis of copper nanoparticles," *Nano,* **7**, p. 1230005, 2012.

[4] M. M. Khalil, E. H. Ismail, and F. El-Magdoub, "Biosynthesis of Au nanoparticles using olive leaf extract: 1st nano updates," *Arab. J. Chem.***5**, pp. 431-37, 2012.

[5] A. Bankar, B. Joshi, A. R. Kumar, and S. Zinjarde, "Banana peel extract mediated synthesis of gold nanoparticles," *Colloids Surf., B.***80**, pp. 45-50, 2010.

[6] P. P. Gan, S. H. Ng, Y. Huang, and S. F. Y. Li, "Green synthesis of gold nanoparticles using palm oil mill effluent (POME): A low-cost and eco-friendly viable approach," *Bioresource Technol.***113**, pp. 132-35, 2012.

[7] T. S. Santra, F.-G. Tseng, and T. K. Barik, "Green biosynthesis of gold nanoparticles and biomedical applications," *American Journal of Nano Research and Application,***2**, pp. 5-12, 2014.

[8] R. Sharma, S. Gulati, and S. Mehta, "Preparation of gold nanoparticles using tea: a green chemistry experiment," *J. Chem. Educ.,* **89**, pp. 1316-18, 2012.
[9] M. Moniruzzaman, K. Nakashima, N. Kamiya, and M. Goto, "Recent advances of enzymatic reactions in ionic liquids," *Biochem. Eng. J.,* **48**, pp. 295-314, 2010.
[10] H. Passos, M. G. Freire, and J. A. Coutinho, "Ionic liquid solutions as extractive solvents for value-added compounds from biomass," *Green Chem.,* **16**, pp. 4786-4815, 2014.
[11] K. P. Charan, N. Pothanagandhi, K. Vijayakrishna, A. Sivaramakrishna, D. Mecerreyes, and B. Sreedhar, "Poly (ionic liquids) as "smart" stabilizers for metal nanoparticles," *Eur. Polym. J.,* **60**, pp. 114-22, 2014.
[12] S. Mohamed, "Oil palm leaf: a new functional food ingredient for health and disease prevention," *J. Food Process Technol.,* **5**, 2014.
[13] J. Anuradha, T. Abbasi, and S. Abbasi, "An eco-friendly method of synthesizing gold nanoparticles using an otherwise worthless weed pistia (Pistia stratiotes L.)," *JAR,* **6**, pp. 711-20, 2014.
[14] S. Bhuvanasree, D. Harini, A. Rajaram, and R. Rajaram, "Rapid synthesis of gold nanoparticles with Cissus quadrangularis extract using microwave irradiation," *Spectrochim. Acta Mol. Biomol. Spectrosc.,* **106**, pp. 190-96, 2013.
[15] K. Sneha, M. Sathishkumar, S. Y. Lee, M. A. Bae, and Y.-S. Yun, "Biosynthesis of Au nanoparticles using cumin seed powder extract," *J. Nanosci. Nanotechnol.,* **11**, pp. 1811-14, 2011.
[16] A. Mishra, M. Kumari, S. Pandey, V. Chaudhry, K. Gupta, and C. Nautiyal, "Biocatalytic and antimicrobial activities of gold nanoparticles synthesized by Trichoderma sp," *Bioresource Technol.,* **166**, pp. 235-42, 2014.
[17] M. Irfan, T. Ahmad, B. Abdullah, S. Bhattacharjee, and M. Moniruzzaman, "Size and Stability Modulation of Ionic Liquid Functionalized Gold Nanoparticles Synthesized Using Elaeis guineensis (Oil Palm) Kernel Extract," *Arab J. Chem.,* 2017.http://dx.doi.org/10.1016/j.arabjc.2017.02.001.
[18] K. Anand, R. Gengan, A. Phulukdaree, and A. Chuturgoon, "Agroforestry waste Moringa oleifera petals mediated green synthesis of gold nanoparticles and their anti-cancer and catalytic activity," *J. Ind. Eng. Chem.,* **21**, pp. 1105-11, 2015.
[19] K. U. Suganya, K. Govindaraju, V. G. Kumar, T. S. Dhas, V. Karthick, G. Singaravelu, *et al.*, "Blue green alga mediated synthesis of gold nanoparticles and its antibacterial efficacy against Gram positive organisms," *Mater. Sci. Eng.,: C,* **47**, pp. 351-356, 2015.
[20] M. Irfan, T. Ahmad, M. M. Moniruzzaman, B. Abdullah, and S. Bhattacharjee, "Ionic Liquid Mediated Biosynthesis of Gold Nanoparticles Using Elaeis Guineensis (Oil Palm) Leaves Extract," *Procedia Eng.,* **148**, pp. 568-72, 2016.

IOP Conf. Series: Materials Science and Engineering **204** (2017) 012003 doi:10.1088/1757-899X/204/1/012003

Electrical and Structural Analysis on the Formation of n-type Junction in Germanium

Umar Abdul Aziz[1], Nur Nadhirah Mohamad Rashid[1], Siti Rahmah Aid[1], Anthony Centeno[1], Hiroshi Ikenoue[2] and Fang Xie[3]

[1]Malaysia-Japan International Institute of Technology, Universiti Teknologi Malaysia, Jalan Sultan Yahya Petra, 54100 Kuala Lumpur, Malaysia
[2]Graduate School of Information Science and Electrical Engineering, Kyushu University, Fukuoka, Japan
[3]Department of Material, Imperial College London, England, United Kingdom

E-mail: sitirahmah.aid@utm.my

Abstract. Germanium (Ge) has re-emerged as a potential candidate to replace silicon (Si) as a substrate, due to its higher carrier mobility properties that are the key point for the realization of devices high drive current. However, the fabrication process of Ge is confronted with many problems such as low dopant electrical activation and the utilization of heavy n-type dopant atoms during ion implantation. These problems result in more damage and defects that can affect dopant activation. This paper reports the electrical and structural analysis on the formation of n-type junction in Ge substrate by ion implantation, followed by excimer laser annealing (ELA) using KrF laser. ELA parameters such as laser fluences were varied from 100 - 2000 mJ/cm^2 and shot number between 1 - 1000 to obtain the optimized parameter of ELA with a high degree of damage and defect removal. Low resistance with a high degree of crystallinity is obtained for the samples annealed with less than five shot number. Higher shot number with high laser fluence, shows a high degree of ablation damage.

1. Introduction

The International Technology Roadmap for Semiconductor (ITRS) has stated that traditional scaling of planar silicon Metal Oxide Semiconductor (Si-MOS) has reached its potential downscaling limit. Further downscaling will lead to the Short Channel Effect (SCE) that will result in an increase of devices leakage current [1–4]. This leakage current will increase the power consumption of devices while reducing the performance of the device. High drive current capability of devices without further downscaling process can be realized by increasing the carrier mobility in the substrate. Germanium (Ge) has emerged as one of the potential candidates to replace Si as a substrate for MOS transistor, due to its higher electrical carrier mobility (3900 cm^2/ V.s for electrons and 1900 cm^2/ V.s for holes) [5]. Furthermore, its similarity with the conventional Si will ease the replacement process in manufacturing lines [6–8].

The fabrication of Ge-based MOS transistor presents several problems associated with poor dopant solubility, low dopant electrical activation level and large dopant diffusion coefficient. All of these problems will limit the high drive current capability of the devices and increase leakage current [9–12]. Moreover, the fabrication of n-MOS transistor is more problematic than p-MOS due to the utilization of

Content from this work may be used under the terms of the Creative Commons Attribution 3.0 licence. Any further distribution of this work must maintain attribution to the author(s) and the title of the work, journal citation and DOI.
Published under licence by IOP Publishing Ltd

IOP Conf. Series: Materials Science and Engineering **204** (2017) 012003 doi:10.1088/1757-899X/204/1/012003

heavy dopant ions i.e. arsenic (As) and antimony (Sb) during the ion implantation process resulting in more damage and defects [12]. The damages and defects will interact with the dopant atoms during thermal annealing and result in electrical deactivation. Therefore, further optimization of thermal annealing is important to achieve high dopant concentration profile in order to realize a higher electrically active carrier concentration as well as damage and defect removal in the substrate after ion implantation process.

Recently, excimer laser annealing (ELA) has gained interest compared to other conventional thermal annealing i.e. Rapid Thermal Annealing (RTA) and Furnace Annealing (FA) since it can achieve higher carrier concentration with a high level of damage/defect removal [13–15]. Due to Ge low melting point, ELA is very efficient for resolidification of ion-implanted Ge substrate at lower laser energy density compared to Si [16].

This paper reports the electrical and structural analysis of n-type junction in Ge substrate by ion implantation process followed by KrF excimer laser. Laser fluences and shot numbers were varied to find the optimized parameter of laser annealing. The annealed sample was experimentally analyzed by I-V characteristics for the resistance value, FESEM for surface images and Raman spectroscopy for the recrystallization. The result of the analysis for the annealed sample is discussed.

2. Experimental Details

Ge (100) wafer was implanted with P at an energy of 20 keV and dose concentration of 3.5×10^{14} cm^{-2}. This combination is important in order to achieve simulated maximum dopant concentration of $1 \times 10^{20} cm^{-3}$ [17]. 7° tilt angle was selected to minimize the channeling effect during the ion implantation process [18–19]. Prior to the thermal annealing process, Ge implanted sample was cleaned using de-ionized water for 1 minutes. Laser thermal annealing was then performed using KrF excimer laser (λ = 248 nm) with a pulse duration of 55 ns in an ambient atmosphere. The laser spot size is 300 µm × 300 µm. Laser fluences were varied from 100 to 2000 mJ/cm^2 and shot numbers varied from 1 to 1000 shots. Current-Voltage (I-V) measurement were taken for each annealed spot. Selected area was measured by Field Emission Scanning Electron Microscope, FESEM to obtain surface morphology of the annealed spots. Recrystallization of annealed sample was then characterized using Raman Spectroscopy.

3. Results and Discussion

Each laser matrix spot (300 µm × 300 µm) was analysed by I-V characteristics in order to measure the resistance of implanted layer. The result of resistance obtained from I-V characteristics is shown in figure 1. Resistance of non-irradiation area is also included as a reference. Stable tendency of resistance with value of less than 1.0×10^3 Ω can only be achieved when the shot number were below five. These low resistance values from I-V characteristics indicated that high level of activation can be obtained after the laser annealing process. This activation takes place by dopant substitution onto the lattice site during the recrystallization process. The level of damage removal can be expected to be higher in these samples. On the other hand, high laser fluence with high shot number show fluctuating resistance. This is due to the ablation damage to the annealed sample as observed from FESEM analysis.

Figure 2 shows the FESEM images of laser annealing spot for various fluence and shot number. Higher laser fluence and shot number results in high degree of ablation damage and surface roughness; as observed from figure 2 (a) with 2000 mJ/cm^2 and 1000 shots. This roughness damage leads to the dopant loss which reduced the concentration of dopants in the substrate. Ablation damage was reduced significantly with reduced laser fluences and shot numbers as observed from figure 2(a-d).

Due to the usage of high laser fluence and shot number, it can be considered that the energy deposition and surface temperature of annealed spot increased [20]. Melted substrate merge together resulting into bump formation on the outer layer of annealed spot and decrease the dopant substrate density; as can be observed from the rough surface of figure 2(a). The decrease in substrate density may results in the dopant loss, which subsequently leads to decrease of the attained dose of implanted dopant near the surface. Hence, the increase in resistance value as shown in figure 1. Bump formation is related to increase in local volume of the substrate, due to increased stresses by laser heating [20]. This can be related to the surface roughness that eventually will affect the stability and accuracy of electrical measurement in the samples (refer to figure 1). Therefore, it is important to select appropriate laser parameter (i.e. laser fluence and shot number) to reduce the laser annealing damages that affect the physical and electrical characteristics of the fabricated sample.

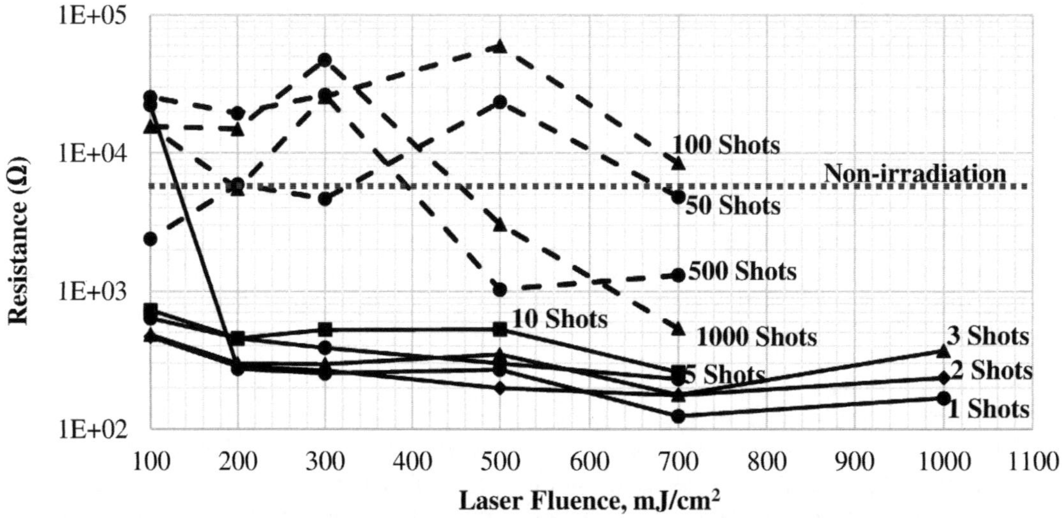

Figure 1 Fluence vs resistance graph obtains from I-V characteristics for ELA center parameter

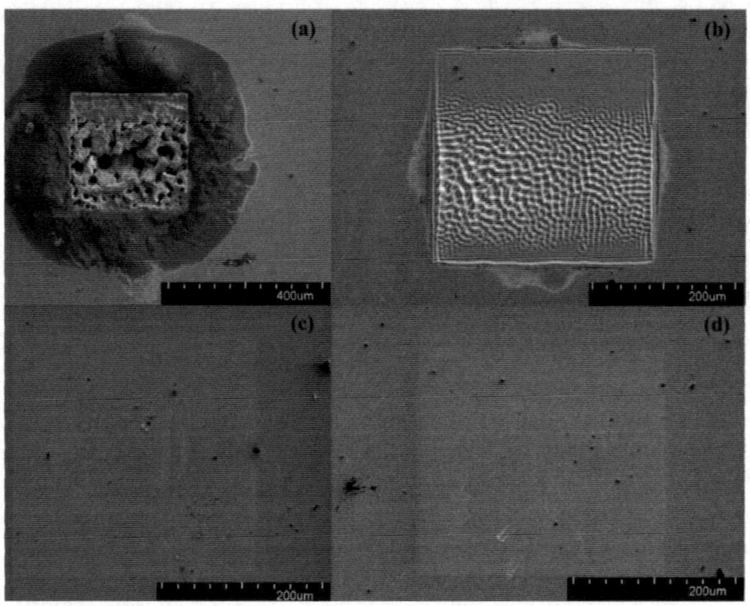

Figure 2 FESEM images of laser annealing spot (300 µm × 300 µm) of difference laser fluence and shot number (a) 2000 mJ/cm^2 with 1000 shots (b) 1000 mJ/cm^2 with 100 shots (c) 500 mJ/cm^2 with 5 shots and (d) 300 mJ/cm^2 with 5 shots

Figure 3 shows the Raman spectroscopy of the samples annealed with shot number below five; with laser fluences from 300 to 2000 mJ/cm^2. The broad peak phase of damage region (i.e. amorphous Ge ,a-Ge), is at 270 cm^{-1} while for crystalline Ge (c-Ge) region is around 300 cm^{-1}. Peak shift of c-Ge samples can

be observed when the ELA were performed with laser fluences above 300 mJ/cm^2; and shot number ranging from two to five. This indicates recrystallization of damage layer has occurred with the degree of crystallinity improved along with the laser fluences. However, ablation damage might occurred in the sample annealed with higher laser fluence. This should be taken into consideration during the parameter selection.
The significant of high degree of crystallinity with lower resistance can be explained by the melt-regrowth model [21]. When a-Ge were completely melted due to ELA, epitaxial regrowth of the melted region occurred, which enable high activation of the dopants. A good resolidification indicates good crystallinity structure that will provide much lower leakage current.

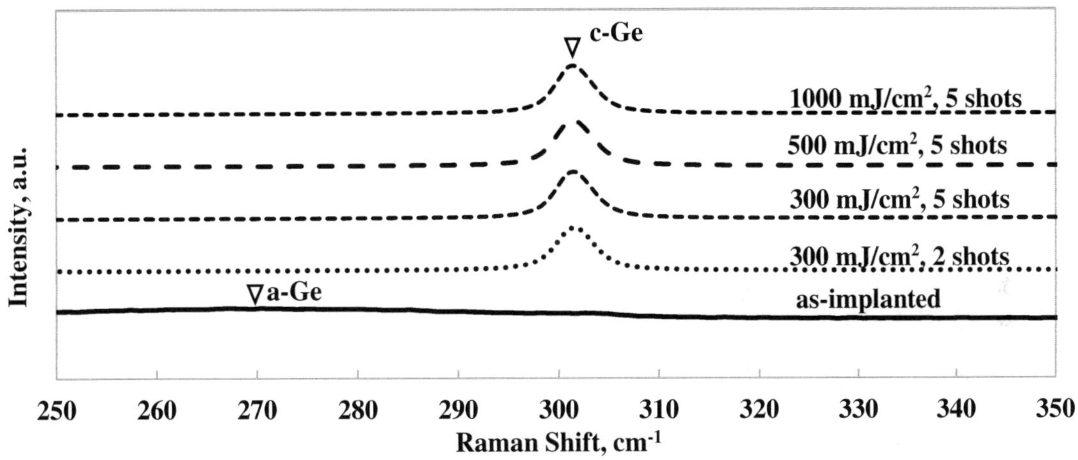

Figure 3 Raman spectra of annealing sample at laser fluence vary from 300 to 1000 mJ/cm^2 with shot number vary from two to five

4. Conclusion

In this work, fabrication of Ge pn junction using KrF excimer laser was demonstrated. Optimization of laser thermal annealing was performed by varying the laser fluences and shots number. The electrical and structural analysis was performed to obtain optimized parameter of laser annealing. It is found that low resistance with a high degree of crystallinity is obtained for the samples annealed with less than five-shot number. Higher shot number with high laser fluence, shows a high degree of ablation damage. For the future works, samples will be fabricated with lower energy fluence (i.e. lower than 500 mJ/cm^2) with shot number below five shots to measure sheet resistance, R_s of implanted layer with higher accuracy. Recrystallization layer will also be quantitatively analysed using Ellipsometry to support Raman spectroscopy analysis.

Acknowledgments

This work is supported by Imperial College Global Engagement Grant, Universiti Teknologi Malaysia International Networking Grant (PY/2015/04262), Takasago Research Grant (OTR) (PY/2015/05503; Vote No: 4B211) and Ministry of Education, Malaysia, Universiti Teknologi Malaysia Research University Grant Tier 1 (PY/2016/06558; Vote No: 14H01). The authors would like to acknowledge the support by Prof. Dr. Satoru Matsumoto and Mr. Akira Suwa (Department of Gigaphoton Next GLP, Kyushu University) for the contributions in this work.

References

[1] Information on ITRS 2010 Report at www.itrs.net
[2] Aid S R, Matsumoto S and Fuse G 2011 *Physica Status Solidi A.* **208** 1646-1651
[3] Aid S R, Matsumoto S, Fuse G, and Sakuragi S 2011 *Physica Status Solidi A.* **208** 2772-2777

IOP Conf. Series: Materials Science and Engineering **204** (2017) 012003 doi:10.1088/1757-899X/204/1/012003

[4] Aid S R, Hara S, Shigenaga Y, Fukaya T, Tanaka Y, Matsumoto S, Fuse G and Sakuragi S 2013 Japanese J. of Appl. Phys. **52** 026501

[5] Sze S M, Ng K K *Physics of Semiconductor Devices* (New York: Wiley- Interscience)

[6] Shang H, Frank M M, Gusev E P, Chu J O, Bedell S W, Guarini K W and Leong M 2006 *IBM J. Res. & Dev.* **50** 337

[7] Lee C H, Nishimura T, Tabata T, Wang S K, Nagashio K, Kita K, and Torium A 2010 *IEDM.* 18.1.1

[8] Simoen E and Vanhellemont J 2009 *Journal of Appl. Phy.* **106** 103516

[9] Thareja G, Liang J, Chopra S, Adams B, Patil N, Cheng S L, Nainani A, Tasyurek E, Kim Y, Moffatt S, Brennan R, McVittie J, Kamins T, Saraswat K and Nishi Y 2010 *IEDM* 10.5.1

[10] Takagi S, Tezuka T, Irisawa T, Nakaharai S, Numata T, Usuda K, Sugiyama N, Shichijo M, Nakane R and Sugahara S 2007 *Solid-State Electron.* **51** 526

[11] Mirabella S, De Salvador D, Napolitani E, Bruno E and Priolo F 2013 *J. of Appl. Phy.* **113** 031101

[12] Kim J, Bedell S W and Sadana D K 2011 *Appl. Phy. Lett.* **98** 082112

[13] Shayesteh M, O' Connell D, Gity F, Stud. Mem, IEEE, Murphy-Armando P, Yu R, Huet K, Toqué-Tresonne I, Cristiano F, Boninelli S, Henrichsen H H, Nielsen P F, Petersen D H and Duffy R 2014 *IEEE Trans. on Elect. Dev.* **61** 4047

[14] Hellings G, Rosseel E, Simoen E, Radisic D, Peterson D H, Hansen O, Nielsen P F, Zschatsch G, Nazir A, Clrysse Vandervorst T W, Hoffmann T Y and De Meyer K 2011 *Elec. Solid-State Lett.* **14** H39-H41

[15] Milazzo R, Napolitani E, Impellizerri G, Fisicaro G, Boninelli S, Cuscuna M, De Salvador D, Mastromatteo M, Italia M, La Magna A, Fortunato G, Priolo F, Privitera V and Carnera A 2014 *J. of Appl. Phy.* **115** 053501

[16] Wang C, Li C, Huang S, Lu W, Yan G, Zhang M, Wu H, Lin G, Wei J, Huang W, Lai H and Chen S 2014 *Appl. Surface Science* **300** 208-212

[17] Aziz U A, Arissa N F, Aid S R, Yahaya H, Centeno A, Matsumoto S, Uedono A and McPhail D 2015 *MJJIC 2015*

[18] Chui C O, Kulig L, Moran J and Tsai W 2005 *Appl. Phys. Lett.* **87** 09190

[19] Ioannou N, Skarlatos D, Tsamis, Krontiras A, Georga S N, Christofi A and McPhail D S 2008 *Appl. Phy. Lett.* **90** 101910

[20] Iqbal M H, Bashir S, Rafique M S, Dawood A, Akram M, Mahmood K, Hayat A, Ahmad R, Hussain T and Mahmood A 2015 *Appl Surf. Sci.* **344** 146-158

[21] Preston J S and van Driel H M 1984 *Phys. Rev B.* **30** 1950–1956

IOP Conf. Series: Materials Science and Engineering **204** (2017) 012004 doi:10.1088/1757-899X/204/1/012004

Low-Temperature Synthesis of Vertically Align ZnO Layer on ITO Glass: The Role of Seed Layer and Hydrothermal Process

Amalia Sholehah[1,2], NurSumiati Achmad[1,2], Arbi Dimyati[3], Yanyan Dwiyanti[2], Tri Partuti[2]

[1] Advanced Material and Tomography Laboratory, Faculty of Engineering, University of Sultan AgengTirtayasa, Jl. JenderalSudirman Km 3, Cilegon – Banten 42435, Indonesia

[2] Department of Metallurgical Engineering, University of Sultan AgengTirtayasa, Jl. Jenderal Sudirman Km 3, Cilegon – Banten 42435, Indonesia

[3]National Nuclear Energy Agency of Indonesia, KawasanPuspiptekSerpong, Gedung71, Tangerang – Banten 15310, Indonesia

E-mail: sitirahmah.aid@utm.my

Abstract. ZnO thin layer has a broad potential application in optoelectronic devices. In the present study, vertically align ZnO layers on ITO glass were synthesized using wet chemical method. The seed layers were prepared using electrodeposition method at 3°C. After that, the growing process was carried out using chemical bath deposition (CBD) at 90°C. To improve the structural property of the ZnO layers, hydrothermal technique was used subsequently. Results showed that seeding layer has a great influence on the physical properties of the ZnO layers. Moreover, hydrothermal process conducted after the ZnO growth can enhance the morphological property of the layers. From the experiments, it is found that the ZnO layers has diameter of ~60 nm with increasing thickness from ~0.8 to 1.2 μm and band-gap energies of ~3.2 eV.

1. Introduction

Zinc oxide (ZnO) is a multi-functional semiconductor with broad applications. Due to its outstanding optoelectronic properties, such as a large excitonic binding energy of 60 meV and a wide direct band gap of 3.37 eV [1,2], ZnO thin film has been considered for optoelectronic device, such as light-emitting diodes (LEDs), laser diodes, photodetectors, and solar cells.

Dye-sensitized solar cell (DSSC) is a third generation solar cell which has been gaining large attentions due to its low-cost fabrication and esthetical benefit. Zinc oxide has been widely used as semiconductor layer in DSSC's photoanode. It is commonly suggested that surface condition of the ZnO nanostructure is a key factor in determining cell performance [3]. The growth pattern and alignment of ZnO layer is important in the semiconductor researches.

In order to obtain good alignment, growing order, coverage, and homogeneity, a seed layer can be prepared prior to the film growing. Among all other methods, such as spin-coating [4], sputtering [5], spray pyrolysis [6], and pulsed laser deposition (PLD) [7], electrochemical deposition (electrodeposition) offers the most efficient route to produce thin, homogenous seed layer, within low temperature. Previous works has revealed that electrodeposition method can generate uniform and

Content from this work may be used under the terms of the Creative Commons Attribution 3.0 licence. Any further distribution of this work must maintain attribution to the author(s) and the title of the work, journal citation and DOI.
Published under licence by IOP Publishing Ltd

IOP Conf. Series: Materials Science and Engineering **204** (2017) 012004 doi:10.1088/1757-899X/204/1/012004

reproducible ZnO layer [1,8]. However, different conditions in deposition process can lead to various morphological and structural modification of ZnO layer [9].

In this study, ZnO were deposited on indium tin oxide glass (ITO) glass using electrodeposition method. The process was carried out using simple two-electrode and low-temperature system. Morphological, structural, and basic optical properties of the ZnO layer were further analyzed.

2. Experimental Design

The seed solution was prepared using zinc nitrate tetrahydrate ($Zn(NO_3)_2.4H_2O$/Zn-nitrate, Merck), heksametilentetraamin ($C_6H_{12}N_4$ / HMTA, Merck). Prior to the seed layer deposition, the indium tin oxide ($InSn_2O_3$, ITO) glass was cleaned in water, acetone, and ethanol with ultrasonic cleaner for 480 seconds in each liquid. After that, the ITO glass was dried and stored in a dry place.

The solution was made by dissolving an equimolar Zn-nitrate and HMTA 0.1 M at 0°C and allowed to stand for 1 hour. The electrodeposition system was constructed using ITO glass as the anode and Cu metal as the cathode. The experiment was conducted using voltage of -1 V for 1 hour in cold bath (3°C), until white, thin seeding layer was seen on top of the ITO glass.

The ZnO layer on ITO glass was grown using chemical bath deposition (CBD) method. The seeded ITO glass was placed vertically in equimolar Zn-nitrate solution and HMTA 0.05 M at 90°C for 3 hours. After that, the ITO glass was washed using distilled water and dried in air.

Hydrothermal treatment was performed by placing the ITO with ZnO layer on top of boiling water in a reactor. The ZnO layer was placed upside down, facing the bottom of the reactor. There are two variations of the process undertaken in the experiment. The first variation is carried out using a hydrothermal reactor in a muffle furnace at a temperature of 150°C for 3 hours at atmospheric pressure. This hydrothermal treatment was further referred by the acronym HT-1. In the second variation, hydrothermal process was carried out used a closed reactor at 100°C for 1 hour under 1 bar nitrogen gas (N_2) pressure. This treatment hereinafter was referred by the acronym HT-2. For each hydrothermal process, the sample collected was further washed with distilled water, dried, and characterized to determine the physical properties.

The morphology of the obtained ZnOnanorods was examined using field-emission scanning electron microscopy (FE-SEM, FEI – Quantana 650), while the crystal's structural information was gained using X-ray diffraction (XRD, Pan Analytical X-Pert Pro). Optical information of the samples were gained using ultraviolet-visible spectrophotometer (UV-VIS Spectrophotometer, Shimadzu 2450) in diffuse reflectance spectroscopy (DRS) mode.

3. Results and Discussions

The morphological image of ZnO layers on ITO glass were shown in figure 1. Most of the ZnO were grown in nanorods structure. However, the as-grown ZnOnanorods in the as-grown layer were not vertically aligned (figure 1.a). This is a typical condition for ZnO layers which are prepared from solution phase. In order to enhance the ZnO layer alignment, an orientation process was conducted afterwards. Using hydrothermal methods, the nanorods were shown to be more vertically oriented (figure 1.b and 1.c) compared to the as-grown ZnO layer. Varied condition of the hydrothermal process has given different results. In HT-1 treatment, the nanorods has the biggest diameter among other variations. This correlates to the high temperature and longer hydrothermal time, which induce the grain growth of ZnOnanorods (figure 1.b). In contrast, HT-2 treatment has given the smallest ZnOnanorods diameter (figure 1.c). When ZnO layer was formed in CBD process, some of the nanorods might be grown together as a bundle. Using the N_2 pressure, the bundle can be separate which results in smaller diameter nanorods.

In the entire process, ZnOnanorods diameter did not change very drastically after the hydrothermal process was conducted. This is caused by the presence of the seed layer, which was formed prior to the growth process of ZnO layers. In this work, electrodeposition has proven to be useful method to produce a good ZnO seed layer.

IOP Conf. Series: Materials Science and Engineering **204** (2017) 012004 doi:10.1088/1757-899X/204/1/012004

Figure 1. Top views of ZnO a) as-grown using CBD method, b) after HT-1 treatment, and c) after HT-2 treatment.

Hydrothermal treatment has proven to not only improve the direction of growth of ZnO, but the density of the ZnO layer as well. Figure 2 shows that the direction of as-grown ZnO which is not entirely vertical, can be corrected after the hydrothermal process is carried out. In figure 2.a, it is shown that some of the ZnOnanorods were grown in a tilted position. However, adding pressure by the hydrothermal process, could force the tilted-grown ZnO to be perpendicular. Thus, the density of ZnO could be enhanced (figure 2.b and 2.c). However, the addition of N_2 has produced the best ZnO layer (figure 2.c). The average diameters and thickness of the ZnO layers is presented in table 1.

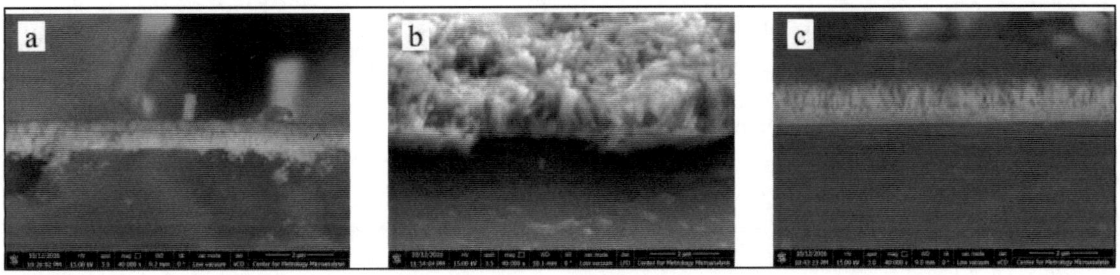

Figure 2. Cross-sectional views of ZnO a) as-grown using CBD method, b) after HT-1 treatment, and c) after HT-2 treatment.

Table 1. Average diameters and thickness of ZnO layers.

ZnOnanorods properties	CBD	HT-1	HT-2
Average diameter (nm)	64.30	68.21	60.44
Thickness (μm)	1.24	0.82	1.20

The structural analysis of ZnO layers were carried out using x-ray diffraction. Figure 3 shows the diffractograms of the layers. All peaks were confirmed with ICSD no 98-016-2843. The highest peaks were found in (002) plane, indicating that at these conditions most of the nanorods were grown vertically along the *c*-axis. The diffractograms shown in figure 3 has good correlation with figure 1 and 2 previously presented.The estimation of the average crystallite size of the wurtzite phase in samples was performed using Scherrer's formula on the diffraction data [10]. The average crystallite diameters of ZnO layers are presented in table 2.

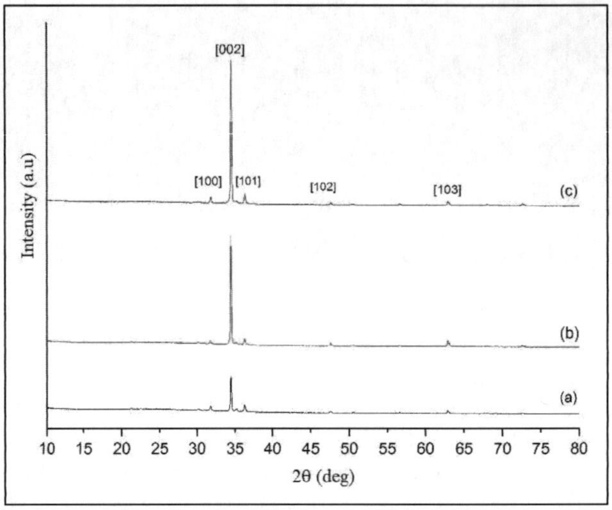

Figure 3. Diffractograms of ZnO a) as-grown using CBD method, b) after HT-1 treatment, and c) after HT-2 treatment.

Table 2. Crystallite size of ZnO layers on ITO.

ZnOnanorods properties	CBD	HT-1	HT-2
Average crystallite size(nm)	61.22	64.77	59.70

The effect of nanostructural characteristics on the optical properties was further investigated by UV-Vis spectroscopy. In figure 4, the absorbance spectra of ZnO layers, showing a strong absorbance in the UV region and high transparency in the visible region. A linear slope of the absorption curve was made on each spectrum in order to estimate the absorption edge. Furthermore, the respective band gap energy, E_g, of ZnOnanorods was analyzed based on these absorption edges using absorbance spectrum fitting (ASF) method [11]. The band-gap energies of ZnO layers are presented in table 3.

Table 3. Band-gap energies of ZnO layers on ITO.

ZnOnanorods properties	CBD	HT-1	HT-2
Band-gap energy (eV)	3.23	3.19	3.25

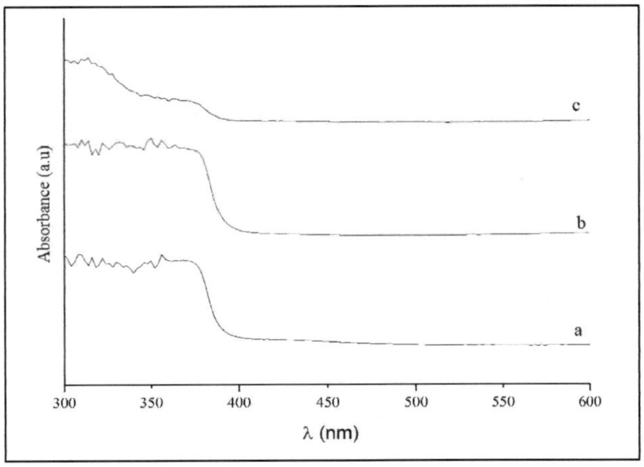

Figure 4. Absorbance spectra of ZnO a) as-grown using CBD method, b) after HT-1 treatment, and c) after HT-2 treatment

All ZnO layers were found to have band-gap energies around 3.2 eV. This can be associated with the presence of seed layers that has been mentioned previously. It can be inferred that the seed layers can maintain not only a good form of the ZnO layers, but the optical property as well. Seed layer can induce the growth of high-quality oriented ZnO film [12]. Hence, to obtain a suitable layer, it is important to investigate the proper condition in seeding process.

4. Conclusions

High-quality oriented ZnO layers were successfully grown on ITO glass. The morphological and optical properties of the layers were determined by the presence of seed layer prior to the ZnO growth process. To enhance the alignment and structural property, hydrothermal process was conducted afterwards. Result showed that the hydrothermal process with N_2 gas pressure has given the best layer, with average diameter, crystallite size, and band-gap energy of 60.44 nm; 59. 70 nm; and 3.25 eV, respectively.

Acknowledgments

This research was conducted under competitive grant from the Ministry of Research, Technology, and Higher Education, Republic of Indonesia, with contract number 056/SP2H/LT/DRPM/II/2016.

References

[1] J. Lu, Y. Sun, M. Zhao, L. Cao, J. Xu, G. He, M. Zhang, and Z. Sun, 2016, *Appl. Surf. Sci.*, **366**, 3, p 348
[2] J. Yang, Y. Wang, J. Kong, H. Jia, and Z. Wang, 2015, *Opt. Mater.*, **46**, p 179
[3] J. Chung, J. Lee, and S. Lim, 2010, *Phys. B Condens. Matter*, **405**, 11, p 2593
[4] D. Berger, E. T. Kubaski, T. Sequinel, R. M. da Silva, S. M. Tebcherani, and J. A. Varela, 2012, *Luminescence*, **28**, 6, pp 942-7
[5] B. Pandey, S. Ghosh, P. Srivastava, D. Kabiraj, T. Shripati, and N. P. Lalla, 2009, *Phys. E*, **41**, p 1164
[6] M. Ardyanian and N. Sedigh, 2012, *J. Phys.*, **78**, 4, p 625
[7] J.-I. Hong, J. Bae, Z. L. Wang, and R. L. Snyder, 2009, *Nanotechnology*, **20**, 8, p 85609
[8] N. Kicir, T. Tuken, O. Erken, C. Gumus, and Y. Ufuktepe, 2016, *Appl. Surf. Sci.*, **377**, p 191
[9] N. Ait Ahmed, H. Hammache, L. Makhloufi, M. Eyraud, S. Sam, A. Keffous, and N. Gabouze, 2015, *Vacuum*, **120**, p 100
[10] Prabhu Y.T., Rao K.V., Kumar V. S. S., and Kumari, B.S., 2014, *World J. Nano Sci. Eng.*, **4**, p 21
[11] Ghobadi, N., 2013, *Int. Nano Lett.*,**3**, 1,p 2
[12] C. Zhang, 2010, *J. Phys. Chem. Solids*, **71**, 3,p 364

IOP Conf. Series: Materials Science and Engineering **204** (2017) 012005 doi:10.1088/1757-899X/204/1/012005

The impact of electron beam irradiation on Low density polyethylene and Ethylene vinyl acetate

Maziyar Sabet[1], Hassan Soleimani[2]

[1]Peroleum and Chemical Engineering Programme, Universiti Teknologi Brunei (UTB),Bandar Seri Begawan, Brunei Darussalam
[2]Faculty of Science and Information Technology, Department of Fundamental and Applied Sciences, Universiti Teknologi PETRONAS (UTP), Bandar Seri Iskandar, Ipoh, Malaysia

E-mail: mazyiar_sabet@yahoo.com, hassan.soleimani@petronas.com.my

Abstract. Improvement of measured gel content, hardness, tensile strength and elongation at break of Ethylene vinyl acetate (EVA) have confirmed positive effect of electron beam irradiation on EVA.Results obtained from both gel content tests show that degree of cross-linking in amorphous regions is dependent on dose. A significant improvement in tensile strength of neat EVA samples is obtained upon electron-beam radiation up to 210 kGy. Similarly, hardness properties of Low-density polyethylene (LDPE) improvewith increasing electron beam irradiation. This article deals with the impacts of electron beam (EB) irradiation on the properties of LDPE and Ethylene-Vinyl Acetate (EVA) as the two common based formulations for wire and cable applications.

1. Introduction

Ethylene-Vinyl Acetate (EVA) copolymers are used extensively in wire and cable industry for making heat shrinkable insulation, semi-conductive insulation jackets, and flame retardant insulation. The ability to accept high filler loadings without embrittlement or loss of mechanical integrity, and ease of cross-linking made it be used extensively forpolymer products [1, 2]. However, unmodified EVA has some shortcomings which can be overcome by formation of intermolecular covalent bonds between polyethylene chains [3, 4]. In this study It was our intention to obtain comprehensive data from effects of absorbed dose on behavior of gel content, tensile strength, elongation at break, hot set, density, melting point, heat of melting, DSC thermograms, thermogravimetry analysis, dielectric constant, surface resistance, volume resistivity, break down voltage, dielectric loss factor, hardness (Rockwell) and heat-deformation percent for ethylene-vinyl acetate (EVA).The original contribution of this paper compared to other papers is that how physical, mechanical, electrical and thermal properties of EVA with 18% vinyl acetate content change when it is exposed against different absorbing dose rate by electron beam which these data are very suitable for insulation of all wire and cable research tests. Likewise, Low-density polyethylene (LDPE) has the largest annual production in world today. Low cost of LDPE, along with its excellent mechanical, processing, and electrical properties, makes it a good choice for many applications [5-8]. However, LDPE has also some shortcomings, such as low softening temperature, cold flow, and a tendency for stress cracking on contact with chemical reagents. These shortcomings can be overcome by formation of intermolecular covalent bonds between polyethylene chains [9-13]. Formation of a network structure changes thermoplastics into materials that no longer melt and will resist temperature-induced deformation or flow. Electron beam (EB) irradiation of polyethylene improves maximum permissible conductor temperature, maximum

Content from this work may be used under the terms of the Creative Commons Attribution 3.0 licence. Any further distribution of this work must maintain attribution to the author(s) and the title of the work, journal citation and DOI.
Published under licence by IOP Publishing Ltd

allowable temperature in case of ''over-current'', soldering and application even at higher its softening point for a long period of time [14-16]. These approaches will be useful to use LDPE for different applications such as wire insulation, shrinkable tubes and packaging films [17].

2. Results and Discussion:

The gel content of irradiated EVA and LDPE were determined to evaluate content of crosslinking produced by irradiation.Figure 1 shows variation of gel content as a function of absorbed dose for electron-beam exposed of EVA and LDPE. Unirradiated EVA and LDPE were found to be completely soluble in hot xylene; however, solubility reduced significantly due to formation of three-dimensional networks in irradiated samples. The gel content of irradiated EVA and LDPEdue to their amorphous structures increased. In general, irradiation below melting point allows for cross-linking reactions to occur within amorphous part of polymer, whereas more radicals are likely to be trapped in crystalline phases [18]. The radical species formed within amorphous regions can be utilized readily in cross-linking reactions, whereas those radicals trapped in crystalline regions have a much lower mobility and a longer lifetime; such radicals can migrate slowly to amorphous region, where they subsequently react [19-25].

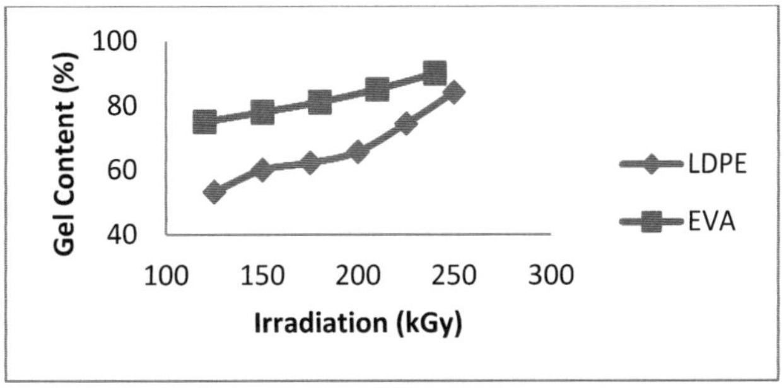

Figure 1 Gel Content of EVA and LDPE as a function of absorbed doses

Figure 2displays variation of density for LDPE and EVA at various absorbed doses. The density of unirradiated EVA and LDPE showed the highest value but increasing irradiation dose decreased density values. This reduction in density is attributed to reduction in crystallinity of LDPE and EVA upon irradiation [1-5].It was reported that increasing irradiation dose made crystalline size smaller and consequently extended amorphous region [6-10].The density of unirradiated EVA shows a slight reduction upon exposing to an electron beam at 120 kGy and then remains almost unchanged when increasing absorbed dose up to 240 kGy. This reduction in density is consistent with reduction in crystallinity of semicrystalline polymers [11-13]. The results obtained from DSC measurements demonstrate slight reductions in melting temperatures of irradiated EVAs. Such slight reductions with increasing dose suggest formation of crystalline regions with slightly reduced sizes. As pointed out earlier, crystalline regions may become slightly impaired due to branching and cross-linking in interfaces between amorphous and crystalline regions [26].

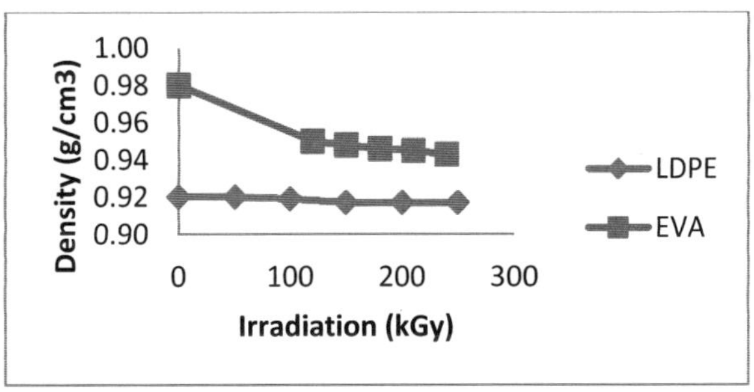

Figure 2 Density of EVA and LDPE as a function of absorbed doses

Figure 3 demonstrates variation of tensile strength of EVA and LDPEat various irradiation doses. The tensile strength of EVA shows a remarkable increase from 13 to about 35MPa. It was shown that upon irradiation tensile strength of EVA increased rapidly up to dose 200 kGy and remained then reduced to 33.5 MPa with further increase of radiation dose to 240 kGy. The reduction is most likely to be caused by chain scission at high dose. It can be understood that for achieving optimum mechanical properties from EVA irradiation dose should not exceed 210 kGy [11].The tensile strength of LDPE increased from 12 to 18.8MPa. Crosslinking and chain scission took place simultaneously over whole range of irradiation doses (0-250 kGy) and it seems that rate of cross-linking was much more dominated [4]. The tighter network in LDPE accounts for high tensile strengths of irradiated LDPE due to restricted movement of molecular chains [5]. Consequently, tensile strength values of LDPE increased after irradiation up to 250 kGy.

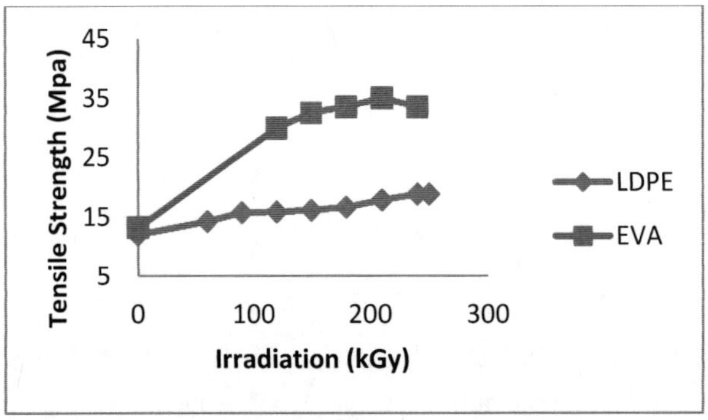

Figure 3 Effect of absorbed dose on tensile strength of EVA and LDPE

Fig. 4 shows relationship between elongation at break and irradiation dose of EVA and LDPE at different irradiation doses from 0 to 250 kGy.The tighter network in EVA samples and high tensile strengths of irradiated samples is due to restricted movement of molecular chains. This causes some losses in elongation of EVA, as shown in Figure 4. As dose increases more cross-links are produced in EVA matrix which prevents structural reorganization during drawing. Therefore by increasing three-dimensional and gel- like structures internal chain mobility and elongation decreases. It is obvious from Figure 4 that elongation at break of EVA was decreased by increasing of absorbed dose [10].Unirradiated LDPE shows the highest value of elongation at break (561%) whereas for all irradiated LDPE due to formation of three dimensional network structures and restricted movement of molecular chain elongation at break decreased sharply to 382% at 250 kGy.Increasing irradiation dose of EVA and LDPEcaused reduction of elongation at break and generated more crosslink bonds in polymeric matrix which prevents structural reorganization during drawing [8, 10-12].

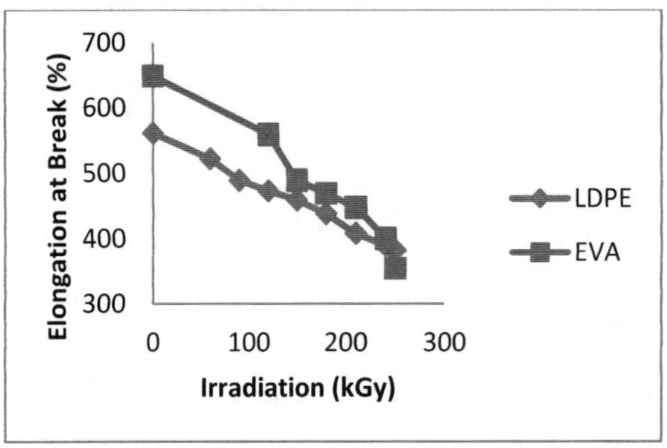

Figure 4 Effect of absorbed dose on Elongation at Break of EVA and LDPE

IOP Conf. Series: Materials Science and Engineering **204** (2017) 012005 doi:10.1088/1757-899X/204/1/012005

Figure 5 shows the dependence of hardness of LDPE and EVA with increasing irradiation dose. It was found that hardness of EVA increased gradually with increase irradiation dose up to 200 kGy which isin agreement with the resultshas been achieved for tensile strength test. Thus, it is evident that enhancement in hardness of EVA is due to formation of radiation cross-links as confirmed by gel content either.Figure 5 also displays that hardness of LDPE increased gradually with increase irradiation dose up to 250 kGy. The hardness value of unirradiated LDPE was measured 32 Rockwell L whereas increasing irradiation dose to 250 kGy increased hardness value to 37.8 Rockwell L. Thus, it confirms that enhancement in hardness of LDPE is due to formation of radiation induced crosslinks as confirmed by gel content and hot set tests. Generally, hardness is referred to resistance of material to local deformation [9], and results proved that irradiated LDPE and EVA were more resistance toward local deformation and increased hardness values.

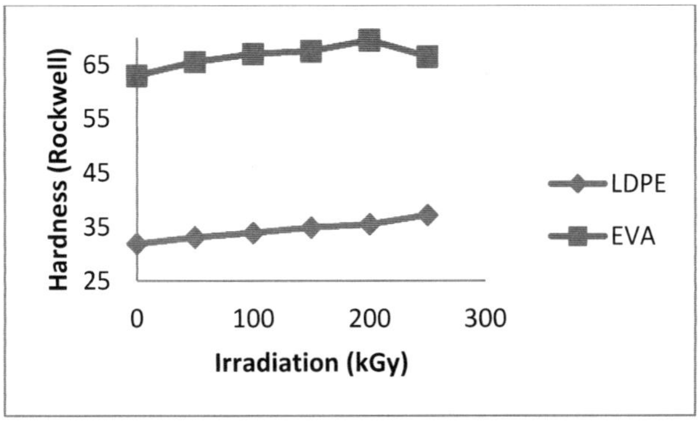

Figure 5 Effect of absorbed dose on Hardness of EVA and LDPE

3. Conclusion

This study investigated effects of electron-beam irradiation on different properties of EVA and LDPE. Cross-link density at a given irradiation dose depends on level of amorphous material in EVA and LDPE. Crystalline regions were only slightly affected by irradiation. Density of EVA decreased slightly with increasing absorbed dose, resulting from slight decreases in crystallinities of EVA. Tensile strength of EVA increased with increasing dose. Cross-linking of EVA increased as dose increased up to 210 kGy. Tensile strength of EVA increased when irradiated up to 210 kGy and decrease with further increase in dose. Hardness of irradiated EVA improved due to formation of radiation cross-links. Deformation of EVA at high temperature reduced with increase of irradiation dose.It has been proved that EVA samples are cross-linked and cross-linking increases with irradiation dose up to about 210 kGy. Mechanical and thermal properties of EVA samples are improved by irradiation and optimum irradiation dose is about 210 kGy. Dielectric constant and loss factor of EVA remain relatively constant and surface and volume resistance of samples were affected significantly by increasing radiation dose. Resistance of EVA increases when radiation dose increases up to 190 kGy. Increasing radiation dose over 190 kGy considerably reduces surface and volume resistance of EVA. Thermal degradation of EVA occurred through two steps as shown by thermogravimetric and derivative curves maximum rate 180 and 310°C. First step involves evolution of acetic acid and second degradation involves decomposition of polyene. Likely, reduction in densities, enthalpies of melting, enthalpies of crystallization, and melting points of LDPE with increase in irradiation dose clearly indicates that crystallinity of LDPE reduces upon irradiation. Thermal degradation of LDPE improves with increase in irradiationdose. A significant improvement in TS of neat LDPE samples obtains upon EB radiation up to 250 kGy with a concomitant decline in elongation at break. Surface resistance, VR, and dielectric strength reveal an approach of maximum value at 250 kGy irradiation dose. However, no considerable changes of dielectric constant and dielectric loss factor are observed with increasing irradiation dose. Enhancement in heat deformation, hardness, and thermal aging properties of LDPE upon EB irradiation suggests that irradiated LDPE is more thermally and mechanically stable than virgin LDPE.

References

[1] M Sabet, A Hassan, C Thevy Ratnam, Properties of ethylene–vinyl acetate filled with metal hydroxide. *Journal of Elastomers and Plastics*. **47**, 88-100, 2015.

[2] D Feldman, Polyolefin, olefin copolymers and polyolefin polyblend nanocomposites, *Journal of Macromolecular Science*, Part A, **53**, 10, 651-658, 2016 .

[3] M Sabet, H Soleimani, Electron-Beam Irradiation of Halogen Free Flame Retardant Polymers for Wire and Cable Applications. *Applied Mechanics and Materials*. **625**, 29-33, 2014.

[4] F Stan, NV Stanciu, C Fetecau, Melt rheological properties of ethylene-vinyl acetate/multi-walled carbon nanotube composites, *Composites Part B: Engineering*, **110**, 20-31, 2017.

[5] M Sabet, H Soleimani, A Hassan, CT Ratnam, Electron Beam Irradiation of LDPE Filled with Calcium Carbonate and Metal Hydroxides. *Journal of Polymer Plastics Technology and Engineering*. **53**, 1362-1366, 2014.

[6] M Sabet, RM Savory, A Hassan, TR Ratnam, Chantara, The Effect of TMPTMA Addition on Electron-beam Irradiated LDPE, EVA and Blend Properties. *International Polymer Processing*. **28**, 386-392, 2013.

[7] Y Pan, L Han, Z Guo, Z Fang, Improving the flame - retardant efficiency of aluminum hydroxide with fullerene for high - density polyethylene, *Journal of Applied Polymer*, **134**, 9, 2017.

[8] S Maziyar, H Azman, TR Chantara, Flammability and thermal characterization of aluminum hydroxide filled with LDPE. *International Polymer Processing*. **28**, 393-397, 2013.

[9] M Entezam, MKR Aghjeh, M Ghaffari, Electron beam irradiation induced compatibilization of immiscible polyethylene/ethylene vinyl acetate (PE/EVA) blends: Mechanical properties and morphology, *Radiation Physics and Chemistry*, **131**, 22-27, 2017

[10] ST Bee, A Hassan, CT Ratnam, TT Tee, Investigation of nano-size montmorillonite on electron beam irradiated flame retardant polyethylene and ethylene vinyl acetate blends, *Nuclear Instruments and Methods in Physics Research* Section B, **299**, 15, 42-50, 2013.

[11] Z Han, Y Wang, W Dong, P Wang, Enhanced fire retardancy of polyethylene/alumina trihydrate composites by graphene nanoplatelets, *Materials Letters*, **128**, 275-278, 2014.

[12] M Sabet, A Hassan, CT Ratnam, Electron-beam irradiation of low density poly ethylene/ ethylene vinyl acetate blends. *Journal of Polymer Engineering*, **33 (2)**, 149-161, 2013.

[13] M Sabet, MSB Anuwar, Calcium Stearate and Alumina Trihydrate Addition of Irradiated LDPE, EVA and Blends with Electron Beam. Journal of Advanced Materials Research, *Applied Mechanics and Materials* **290**, 31-37, 2013.

[14] M Sabet, A Hassan, CT Ratnam, Effect of zinc borate on flammability/thermal properties of ethylene vinyl acetate filled with metal hydroxides. *Journal of Reinforced Plastics and Composites*, *Journal of Reinforced Plastics and Composites* **32 (15)**, 1122-1128, 2013.

[15] SR Chowdhury, BK Sharma, PA Mahanwar, Tensile, flexural and morphological properties of electron beam-crosslinked LDPE–EPDM blends, Plastics, Rubber and Composites, **44**, 10, 440-448, 2015.

[16] M Sabet, H Soleimani, A Hassan, CT Ratnam, Effects of addition calcium stearate and metal hydroxides on the properties of irradiated LDPE/EVA blends. *Journal of Polymer Engineering*, 1-7, 2013.

[17] M Sabet, A Hassan, CT Ratnam, Properties of ethylene–vinyl acetate filled with metal hydroxide. *Journal of Elastomers and Plastics*, June 25, 2013.

[18] ST Bee, LT Sin, CT Ratnam, KJS Haraveen, Investigation of the interaction of copper (II) oxide and electron beam irradiation crosslinkable, polyethylene, *Nuclear Instruments and Methods in Physics Research* Section B, **360**, 36-45, 2015.

[19] M Sabet, A Hassan, CT Ratnam, Mechanical, electrical and thermal properties of irradiated low density poly Ethylene by electron-beam. *Journal of Polymer Bulletin*, **68**, 2323-2339, 2012.

[20] SR Chowdhury, BK Sharma, Vinyl acetate content and electron beam irradiation directed alteration of structure, morphology, and associated properties of EVA/EPDM blends, *Journal of Applied Polymer*, **133**, 21, 2016.

[21] M Sabet, A Hassan, CT Ratnam, Electron beam irradiation of low-density polyethylene filled with metal hydroxides for wire and cable applications. *Journal of Polymer Bulletin*, 23 June, **69**, 1103-1114, 2012.

[22] M Sabet, A Hassan, CT Ratnam, Electron beam irradiation of low density poly ethylene/ ethylene vinyl acetate filled with metal hydroxides for wire and cable applications. *Journal of Polymer Degradation and Stability*, **97**, 1432–1437, 2012.

[23] M Sabet, A Hassan, MU Wahit, CT Ratnam, Mechanical, electrical and thermal properties of irradiated ethylene vinyl acetate by electron beam. *Journal of Polymer-Plastics Technology and Engineering*, Jan.-2010.

[24] S Dadbin, M Frounchi, M Sabet, Studies on properties and structure of E-beam cross-linked LDPE/EVA blends. *Polymer International*, John Willey, **54**, 686-691, 2005.

[25] M. Sabet, H. Soleimani, H., Seyednooroldin, Properties and characterization of ethylene-vinyl acetate filled with carbon nanotube, *Polymer Bulletin*, **73**, 2, 419-434, 2016.

[26] M. Sabet, H. Soleimani, H., Seyednooroldin. Preparation and Characterization of LDPE/CNT, *International Journal of Advances in Science Engineering and Technology*, **4**, 2, 154-160, 2016.

IOP Conf. Series: Materials Science and Engineering **204** (2017) 012006 doi:10.1088/1757-899X/204/1/012006

The structural, morphological and magnetic properties of Ag-doped $La_{0.8}Ca_{0.2-x}Ag_xMnO_3$ synthesized by sol-gel method

F Ruli and B Kurniawan
Department of Physics, Universitas Indonesia, Depok, Indonesia

E-mail: bkuru@fisika.ui.ac.id

Abstract. In this paper, The authors report the structural, morphological and magnetic properties of $La_{0.8}Ca_{0.2-x}Ag_xMnO_3$(x = 0, 0.03) prepared by sol-gel method. The X-ray diffraction (XRD) patterns of samples reveal an orthorhombic perovskite structure with a space group Pnma. The lattice parameters, Mn-O bond length and unit cell volume increase with doping Ag whereas Mn-O-Mn angle decrease. The scanning electron microscopy (SEM) images show that$La_{0.8}Ca_{0.2}MnO_3$sample more porous than $La_{0.8}Ca_{0.17}Ag_{0.03}MnO_3$sample. Analysis using energy dispersive X-ray (EDX) confirms that there is no significant difference between designated and measured composition. M-H loop hysteresis indicate both sampleare paramagnetic materials which $La_{0.8}Ca_{0.17}Ag_{0.03}MnO_3$sample has higher susceptibility than the others.

1. Introduction

The perovskite manganite compounds with general formula $La_{1-x}A_xMnO_3$ (A is a divalent or a monovalent atom) have been intensively studied due to their interplay between structural, electrical and magnetic properties[1,2]. $LaMnO_3$ is an antiferromagnetic and an insulator materials[3]. Subtitution of La site with divalent ions leads to oxidize a Mn^{3+} ion to a Mn^{4+} ion and results in ferromagnetic state and metallic conductivity when the doping levels are sufficiently high[4,5]. Subtitution of La site with monovalent ions oxidizes Mn^{3+} twice as many as divalent ions [6].Relation between electrical and magnetic properties was initially explained by Zener double exchange interaction, hopping electron from Mn^{3+} ions to Mn^{4+} ions through O^{2-} ions[7,8]. Hence, Mixed valence Mn^{3+}/Mn^{4+} ions play a very important role in the electric and magnetic properties of lanthanum manganites [4].

The electrical and magnetic properties are also affected by the avarage A-site cation radius $< r_A>$ and the A-site cation size mismatch [3,9]. Those induce distortion of MnO_6 octahedron thereby change the the distance and angle between Mn-O-Mn ions [10]. Moreover, a strong electron-phonon interaction arising from the Jahn-Teller splitting of the outer Mn d level [11] and the oxygen deficiency [12,13] play a role in the electrical and magnetic properties of lanthanum manganites.

In this present work, the authors partially subtituted divalent ion Ca with monovalent ionAg at perovskite manganite $La_{0.8}Ca_{0.2}MnO_3$and studied the structural, morphological and magnetic properties of that perovskitemananite. Besides having different valence, ionic radii of Ca^{2+} (1.18 Å)are smaller that of Ag^+ (1.28 Å)[14].

2. Experiments

Material $La_{0.8}Ca_{0.2-x}Ag_xMnO_3$ (x = 0 and 0.03) was synthesized by sol-gel method using La_2O_3, $Ca(NO_3)_2.4H_2O$, $AgNO_3$, $Mn(NO_3)_2.4H_2O$ as the precursor materials. La_2O_3 was dissolved in nitric

Content from this work may be used under the terms of the Creative Commons Attribution 3.0 licence. Any further distribution of this work must maintain attribution to the author(s) and the title of the work, journal citation and DOI.
Published under licence by IOP Publishing Ltd

acid and all metal nitrate was dissolved in distilled water. All solution was mixed to form a solution. Citric acid was added to the solution with molar ratio of 1.2 between citric acid and metal nitrate. the pH 7 of solution was achieved by adding amonia solution while stirring the solution at temperature of 80 °C to form gel. The gel was heated at 120 °C for dehidration. It was burned at 600 °C for 5 h to remove produced organic materials during the chemical reaction. The materials obtained from this process was ground by using an agate mortar. The powders were pressed into pellets and then heated at 1000 °C for 24 h.

The structural analysis of the sample was determined by X-ray diffractometer (XRD) with CuKα radiation. The morphological properties were investigated by using scanning electron microscopy (SEM) andthe composition of elements was carriedout using energy dispersive X-ray spectroscopy (EDX). The magnetic properties of samples was determined by Vibrating Sample Magnetometer (VSM).

3. Results and Discussion

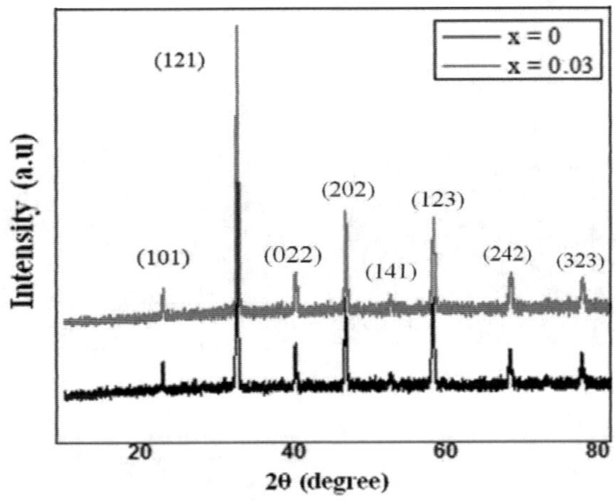

Figure 1. XRD pattern of $La_{0.8}Ca_{0.2-x}Ag_xMnO_3$ (x = 0 and 0.07)

Tabel 1. Lattice parameters (a, b, c), unit cell volume (V), Mn-O bond length, Mn-O-Mn angle, tolerance factor (t) and goodness of fit of $La_{0.8}Ca_{0.2-x}Ag_xMnO_3$ (x = 0 and 0.03)

Sample	$La_{0.8}Ca_{0.2}MnO_3$	$La_{0.8}Ca_{0.17}Ag_{0.03}MnO_3$
a (Å)	5.4763	5.4767
b (Å)	7.7440	7.7458
c (Å)	5.5022	5.5038
Volume (Å^3)	233.34	233.48
$d_{Mn\text{-}O1}$ (Å)	1.9674	1.9677
$\Theta_{Mn\text{-}O1\text{-}Mn}$ (^{o})	161.086	161.085
$d_{Mn\text{-}O2}$ (Å)	1.9654	1.9658
$\Theta_{Mn\text{-}O2\text{-}Mn}$ (^{o})	160.166	160.165
Tolerance factor < t >	0.9249	0.9259
Goodness of fit	1.0898	1.0860

The XRD patterns of $La_{0.8}Ca_{0.2-x}Ag_xMnO_3$samples are shown in figure 1. The refinement results using rietveld method show that the crystal structure of the samples are an orthorhombic perovskite structure with space group Pnma. Goldschmidt's tolerance factor was calculated to know the stability of crystal structure. Tolerance factor (t) was calculated by using equation :

$$t = \frac{r_A + r_O}{\sqrt{2}(r_B + r_O)} \quad (1)$$

where r_A, r_B, and r_O are the radii of A^{3+}, B^{3+}, and O^{2-} ions in ABO_3 perovskite, respectively. Tolerance factor calculation results confirm that the stable crystal structure is orthorhombic perovskite structure ($0.89 \leq t \leq 0.96$) [9]. As shown in table 1, The lattice parameters and unit cell volume of $La_{0.8}Ca_{0.17}Ag_{0.03}MnO_3$samples are larger than $La_{0.8}Ca_{0.2}MnO_3$samples due to ionic radius of Ag^+ (1.28 Å) larger than ionic radius of Ca^{2+} (1.18 Å) [14]. Our results conform with those obtained by Debnath et al [14]. Subtitution of partial Ca site with Ag causes increment of Mn-O bond length and decrement of Mn-O-Mn angle.

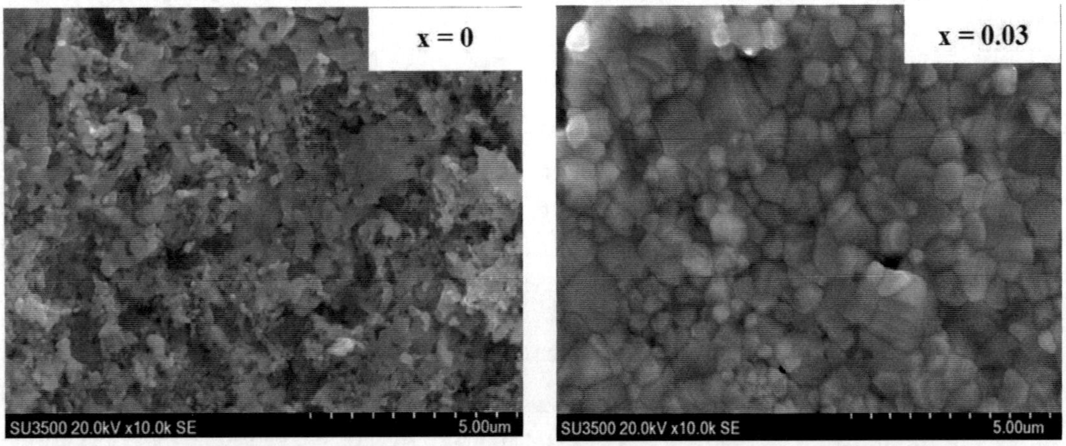

Figure 2. SEM images of $La_{0.8}Ca_{0.2-x}Ag_xMnO_3$samples ($x = 0$ and $x = 0.03$)

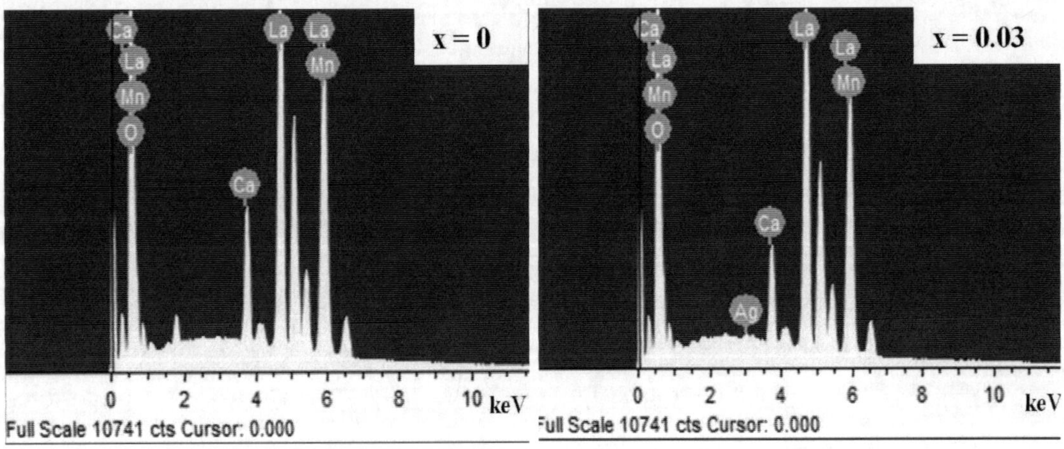

Figure 3. EDX analysis of $La_{0.8}Ca_{0.2-x}Ag_xMnO_3$samples ($x = 0$ and $x = 0.03$)

The SEM image of $La_{0.8}Ca_{0.2-x}Ag_xMnO_3$ samples are presented in figure 2. The $La_{0.8}Ca_{0.2}MnO_3$sample tends to agglomerate. The $La_{0.8}Ca_{0.2}MnO_3$sample is more porous than $La_{0.8}Ca_{0.17}Ag_{0.03}MnO_3$sample. The grain size of $La_{0.8}Ca_{0.17}Ag_{0.03}MnO_3$sample ranges 0.2-1.7 μm. Figure 3shows the results of Energy Dispersive X-ray (EDX) analyses spectrum of $La_{0.8}Ca_{0.2-x}Ag_xMnO_3$. The results show that there is no quantifiable loss of each elements. The compositions of each elements of $La_{0.8}Ca_{0.2-x}Ag_xMnO_3$are presented in table 2.There is a slight difference in quantitative value between designated composition and measured composition as the EDX method is semi-quantitative analysis.

Table 2. Compositional analysis of $La_{0.8}Ca_{0.2-x}Ag_xMnO_3$samples ($x = 0$ and $x = 0.03$)

Doping concentration	Element composition	Designated composition (at. %)	Measured composition (at. %)
	La	16	16.03
	Ca	4	3.71
x = 0	Ag	0	0
	Mn	20	17.11
	O	60	63.14
	La	16	14.75
	Ca	3.4	3.06
x = 0.03	Ag	0.6	0.12
	Mn	20	16.84
	O	60	65.23

Figure 4 shows the M-H loop hysteresis of $La_{0.8}Ca_{0.2-x}Ag_xMnO_3$samples for applied magnetic field 0-2 T. Magnetization for both sample continuously increases with increasing magnetic field indicating that both samples are paramagnetic material at room temperature. The average mass susceptibility of $La_{0.8}Ca_{0.2}MnO_3$ sample is 0.386 emu/g.kOe. This result is comparable with those obtained by Fatnassi et al [15]. The average mass susceptibility of $La_{0.8}Ca_{0.17}Ag_{0.03}MnO_3$ sample is larger than that of $La_{0.8}Ca_{0.2}MnO_3$ sample at room temperature due to its higher ratio in mixed valence Mn [16].

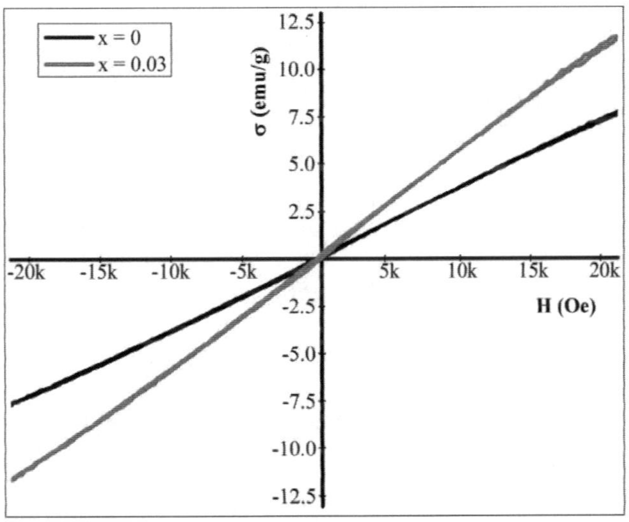

Figure 4. M-H loop hysteresis of $La_{0.8}Ca_{0.2-x}Ag_xMnO_3$samples

4. Summary

The structural, morphological and magnetic properties of $La_{0.8}Ca_{0.2-x}Ag_xMnO_3$samples (x = 0, 0.03) prepared by sol-gel method were investigated. The samples were calcinated at 600 °C and sintered at 1000 °C for 24 h. Both samples were indexed with orthorhombic structure with space group Pnma. The lattice parameters, Mn-O bond length and unit cell volume increase with doping Ag whereas the Mn-O-Mn angle decrease. The $La_{0.8}Ca_{0.2}MnO_3$sample is more porous than $La_{0.8}Ca_{0.17}Ag_{0.03}MnO_3$sample.There is no quantifiable loss of each sample element. $La_{0.8}Ca_{0.2-x}Ag_xMnO_3$samples are paramagnetic materials at room temperature.

Acknowledgements

This work was supported in part by research grant of Ministry of Research, Technology and Higher Education (contract no. 2699/UN.2.R3.1/HKP.05.00/2017) and Indonesia Endowment Fund for Education (LPDP).

References

[1] Rao C and Mahesh R 1998 Giant magnetoresistance, charge ordering and other novel properties of perovskite manganates *J. Phys. Chem. Solids* **3697** pp 487–501

[2] Radaelli P G, Cox D E, Marezio M, Cheong S W, Schiffer P E and Ramirez A P 1995 Simultaneous structural, magnetic, and electronic transitions in $La_{1-x}Ca_xMnO_3$ with x=0.25 and 0.50 *Phys. Rev. Lett.* **75** pp 4488–4491

[3] Venkataiah G, Prasad V and Venugopal Reddy P 2007 Influence of A-site cation mismatch on structural, magnetic and electrical properties of lanthanum manganites *J. Alloys Compd.* **429** pp 1–9

[4] Jonker G H and Van Santen J H 1950 Ferromagnetic compounds of manganese with perovskite structure *Physica* **16** pp 337–349

[5] Schiffer P, Ramirez A P, Bao W and Cheong S W 1995 Low temperature magnetoresistance and the magnetic phase diagram of $La_{1-x}Ca_xMnO_3$*Phys. Rev. Lett.* **75** pp 3336–3339

[6] Ye S L, Song W H, Dai J M, Wang K Y, Wang S G, Zhang C L, Du J J, Sun Y P and Fang J 2002 Effect of Ag substitution on the transport property and magnetoresistance of $LaMnO_3$*J. Magn. Magn. Mater.* **248** pp 26–33

[7] Zener C 1951 Interaction between the d-shell in the Transition Metals. II. Ferromagnetic Compounds of Manganese with Perovskite Structure *Phys. Rev.* **82** pp 440-444

[8] Anderson P W and Hasegawa H 1955 Consideration on Double Exchange *Phys. Rev.***100**675-681

[9] Hcini S, Zemni S, Triki A, Rahmouni H and Boudard M 2011 Size mismatch, grain boundary and bandwidth effects on structural, magnetic and electrical properties of $Pr_{0.67}Ba_{0.33}MnO_3$ and $Pr_{0.67}Sr_{0.33}MnO_3$ perovskites *J. Alloys Compd.* **509** pp 1394–1400

[10] Manjunatha S O, Rao A and Okram G S 2015 Investigation on structural, magneto-transport, magnetic and thermal properties of $La_{0.8}Ca_{0.2-x}Ba_xMnO_3$ ($0 \leq x \leq 0.2$) manganites *J. Alloys Compd.* **640** pp 154–161

[11] Millis A J, Littlewood P B and Shraiman B I 1995 Double exchange alone does not explain the resistivity of $La_{1-x}Sr_xMnO_3$*Phys. Rev. Lett.***74** pp 5144–5147

[12] Khlifi M, Bejar M, El Sadek O, Dhahri E, Ahmed M A and Hlil E K 2011 Structural, magnetic and magnetocaloric properties of the lanthanum deficient in $La_{0.8}Ca_{0.2-x}\square_xMnO_3$ (x = 0-0.20) manganites oxides *J. Alloys Compd.***509** pp 7410–7415

[13] Khlifi M, Dhahri E and Hlil E K 2014 Magnetic, magnetocaloric, magnetotransport and magnetoresistance properties of calcium deficient manganites $La_{0.8}Ca_{0.2-x}\square_xMnO_3$ post-annealed at 800 C *J. Alloys Compd.***587** pp 771–777

[14] Debnath J C, Zeng R, Kim J H and Dou S X 2011 Large magnetic entropy change near room temperature in $La_{0.7}(Ca_{0.27}Ag_{0.03})MnO_3$ perovskite *J. Alloys Compd.***509** pp 3699–36704

[15] Fatnassi D, Rehspringer J L, Hlil E K, Niznansky D, Ellouze M and Elhalouani F 2015 Structural and magnetic properties of nanosized $La_{0.8}Ca_{0.2}Mn_{1-x}Fe_xO_3$ particles ($0 \leq x \leq 0.2$) prepared by sol–gel method *J. Supercond. Nov. Magn.***28** pp 2401–2408

[16] Choura Maatar S, M'nassri R, Cheikhrouhou Koubaa W, Koubaa M and Cheikhrouhou A 2015 Structural, magnetic and magnetocaloric properties of $La_{0.8}Ca_{0.2-x}Na_xMnO_3$ manganites ($0 \leq x \leq 0.2$) *J. Solid State Chem.***225** pp 83–88

IOP Conf. Series: Materials Science and Engineering **204** (2017) 012007 doi:10.1088/1757-899X/204/1/012007

Numerical Study of Two-Phase Flow in Micro-/Nanobubble Generating Pump

Hilman Syaeful Alam[1], Bahrudin[1] and Anto Tri Sugiarto[1]

[1]Technical Implementation Unit for Instrumentation Development, Indonesian Institute of Sciences, Jl. SangkuriangKomplek LIPI Gedung 30 Bandung, 40135, Indonesia.

E-mail: hilm003@lipi.go.id, hilman.syaeful.alam@lipi.go.id,bahr003@lipi.go.id,anto001@lipi.go.id

Abstract. Gas-liquid mixing pump is one of the multiphase flow problem in industrial applications as a micro-/nanobubble generator. However, very few report that studied the two-phase flow for application of microbubble generation because of the analysis complexity. In this paper, a steady state numerical simulation of gas-liquid two-phase flow in the gas-liquid mixing pump was employed to predict a performance and characteristic of fluid flow. Based on simulation results, it is demonstrated that the pump can work in self suction, and generates a vortex flow pattern at every stage of the impeller as regenerative. Performance pump of the numerical simulation is slightly higher than the design specifications Because of mechanical and volume losses was neglected. However, the evaluation method and simulation results from this work can be used as a reference for the design and improvement of the gas-liquid mixing pump.

1. Introduction

Multiphase flow is the most widely applied flow conditions used in various industrial applications. However, a reliable analysis tool for analyzing the problems of multiphase flow is still limited because of the analysis complexity. With the rapid development in the field of computer technology, multiphase flow problems can be solved by numerical analysis based on the computational fluid dynamic (CFD), therefore, it will be significantly reduced the time and cost in the design and production compared to experimental based approach [1]. CFD simulation has been used in previous studies to analyze the problems of multiphase flow, such as: multiphase pump-mixer [2], self-priming pump [3], biodiesel reactor [4], airlift reactor [5], mixed flow pump [6], gas-liquid multiphase scroll pump [7], pump-mix mixer [8] and others.

Gas-liquid mixing pump is one of multiphase flow applications in industries that used for mixing gas and liquid directly as an ultrafine bubble generator. Gas mixed with the liquid in the pump is formed into an ultrafine bubbles with a size <50 um (microbubbles) or <200 nm (nanobubbles), so the gas will easily dissolve in the liquid. In general, this type of generator using regenerative pump that has very simple and inexpensive construction [9]. The fluid enters a set of radial impeller teeth the which guides the fluid from the inlet to the outlet, both separated by about 3000 annulus chamber producing a torroidal motion also known as regenerative flow [10]. Therefore, a regenerative pump can achieve lower mass fluxes but higher-pressure differences than other pumps at the same circumferential velocity [11]. Vortex flow pattern in regenerative pump impact on regenerative effects or impulses

Content from this work may be used under the terms of the Creative Commons Attribution 3.0 licence. Any further distribution of this work must maintain attribution to the author(s) and the title of the work, journal citation and DOI.
Published under licence by IOP Publishing Ltd

that recur the fluid particles at every level of vanes and volute pump channels.Gas fine bubbles would be obtained at the outlet of the impeller due to the super-saturated condition between gas and fluid.
Design of microbubble generator still remains a challenge in generating an efficient design. Very few reports that studied the pattern of two-phase flow in a gas-liquid mixing pump for applications of microbubble generation. In recent years, much more research is carried out experimentally that require a huge costs and a long time. In this paper, the gas-liquid mixing pump was analyzed using CFD approach based on gas-liquid two-phase flow to predict the behavior and pattern of a fluid flow as reference in the design process.

2. Materials and Methods

The main part of gas-liquid mixing pump consists of impeller and housing pump. There are two suctions or inlet fluids of mixing pump, i.e. gas and liquid which is mixed by the impeller component. The design parameter of gas liquid mixing pump was mass flow of 2.2 m3/h, head of 20 m, rotation speed of 3000 r/min, and power of 0.55 kW.

2.1. Computational domains and meshing

According to Huang et al. [3], the conservation equations for each phase at a gas-liquid mixing pump are derived to obtain a set of equations with regard to momentum, continuity and energy. For CFD simulation in a rotary pump, k-e turbulence model and Eulerian multiphase approach had a satisfactory agreement between the experimental result and simulation. Therefore, their model were also used in this case and chosen to close the Reynolds average Navier-Stokes (RANS) equations.
3D model of gas-liquid mixing pump was built using ANSYS Design Modeler was imported into the ANSYS CFX software to generate the computational domain in CFD simulation, as shown in Figure 1(a). The 3-D computational domains of gas liquid mixing pump consist of solid part i.e. impeller and fluid part i.e. volute pump. The volute pump was connected with the liquid and gas suction inlet and pump outlet. The impeller was located in volute pump where the rotation was coupled by the electric motor shaft. The solid and fluid parts were meshed using tetrahedral elements to adjust to the complex geometry of each part. The finer the size of the elements then the numerical accuracy will increase but will also increase the computational time of the simulation. The mesh generation of solid part or impeller produces 50,736 elements and 10,890 nodes as shown in Figure 1(b), while fluid part produces 971,284 cells and 190,944 nodes as shown in Figure 1(c). Finally there were totally 1,022,020 cells and 201,834 nodes. According to Huang et al [3], when cell number achieved 1 million, the variation of pump head under steady simulation could be within 1%.

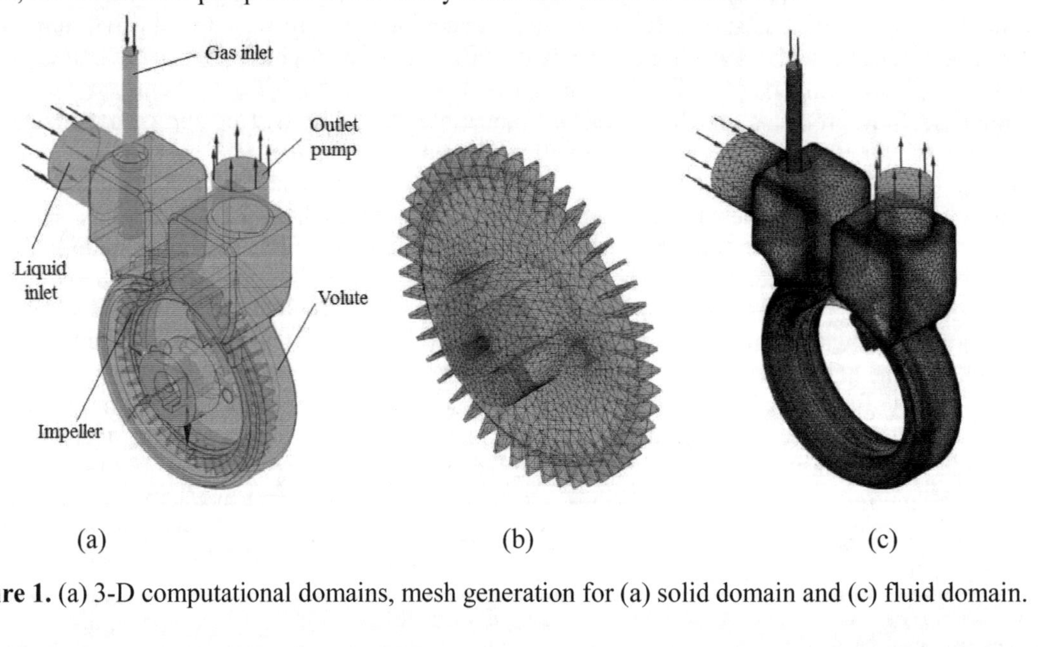

Figure 1. (a) 3-D computational domains, mesh generation for (a) solid domain and (c) fluid domain.

IOP Conf. Series: Materials Science and Engineering **204** (2017) 012007 doi:10.1088/1757-899X/204/1/012007

2.2. Boundary conditions

The next stage of CFD simulation was to set the boundary condition. Water and air at 25 °C were set as the liquid-phase and the gas-phase respectively. The design pump head was used to set the pressure at the inlet and the outlet of gas liquid mixing pump. If the inlet water pressure is set at 1.0 atm, the pressure at the pump outlet is set around 3.0 atm for the pump head of 20 m. Gas liquid mixing pump is designed to be self-suction gas, so that the air inlet pressure is set at 1.0 atm. According to the governing equations in 2.1, the two-phase flow and the standard k-e turbulence models were selected in the governing equations and the steady state simulation was selected in the analysis type. The impeller part was chosen as a continuous solid with aluminum material and rotates on the X coordinate with the rotation of 3000 rpm. The second order upwind scheme was set on the discrete scheme of two phase flow model, the turbulence kinetic energy and the turbulent dissipation rate.

3. Results and Discussion

The distribution of gas liquid two-phase flow on the gas liquid mixing pump as the water volume fraction can be seen in Figure 2(a). The void fraction, $\alpha = 1$ means pure water and $\alpha = 0$ means pure liquid. There is an interesting phenomenon in the analysis results of the figure which are the following: 1. The gas in the gas suction pipe seems inhaled into a liquid storage chamber and impeller indicating that the pump is worked by self-suction mechanisms. 2. The gas is concentrated along the impeller blade and almost uniform from a suction blade to the blade outlet with a concentration above of 75%, it shows if the impeller works well in mixing the gas with liquid. 3. Distribution of gas around the outlet pump looks almost evenly with the liquid gas concentration average was 70%.

Figure 2(b) shows the streamline of the liquid superficial velocity where in the impeller area, it seems that the fluid velocity is lower than the inlet and outlet areas, because the kinetics energy is converted to the pressure energy. Vortex flow pattern is formed in the impeller area which give regenerative effects or impulses that repeatedly on the fluid particles at every level of the vanes and volute pump channels. The magnitude of the fluid velocity is below 18 m/s. Figure 2(c) shows the pressure distribution in the gas-liquid mixing pump on vertical planes. From these images, the pressure at the pump is gradually increased from the blade suction to a peak at the pump outlet. This suggests that the kinetic energy of the blade rotation is gradually converted into pressure energy at each stage of the impeller blade. With a vortex flow pattern and pressure in each stage of the blade, then the fine gas bubbles will be formed at the outlet of the impeller due to the super-saturated condition between gas and fluid. The pressure in the pump outlet was an average of 0.323 MPa and reach a maximum of 0.395 Mpa. It is higher than the design specifications, because CFD simulations did not take into account the factor of both the mechanical and the volume losses caused by bearings and seals.

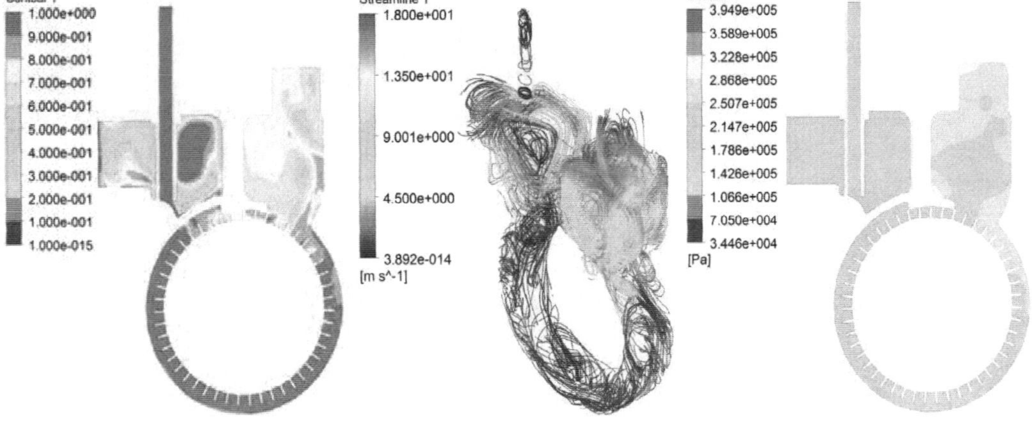

(a) (b) (c)

Figure 2. (a) Gas-liquid two phase-countours on vertical plane (b) Streamline of the liquid superficial velocity, (c)pressure distribution in gas-liquid mixing pump on vertical planes.

4. Conclusion

CFD simulation is able to display the characteristics of gas-liquid two-phase flow on micro-/nanobubble pump. Based on the analysis, the pump can self-suction work, where the gas is successfully sucked into the impeller. The gas is concentrated along the impeller blade uniformly from the suction to the impeller outlet. From the results of streamline analysis, vortex flow pattern proved to be formed in each stage of the impeller that may impact on regenerative effects or impulses thus increasing the pressure in regenerative required to produce ultrafine bubbles at the impeller outlet. The fluid pressure of the numerical simulation results are higher than the design specifications, because both the mechanical and the volume losses caused by bearings and seals was neglected.

Acknowledgement

This work was financed by Kegiatan Unggulan, Indonesian Institute of Sciences (LIPI). The computational facilities were extended by the Research Center of Electrical Power and Mechatronics, Indonesian Institute of Sciences, which is thankfully acknowledged.

References

[1] C. Li, G. H. Yeoh, S. C. P. Cheung and J. Tu, "Modelling Horizontal Gas-Liquid Flow using Averaged Bubble Number Density Approach", *Journal of Computational Multiphase Flows*, **2(2)**, pp. 89-99, 2010.

[2] M. V Tabib, G. Lane, W. Yang, and M. P. Schwarz, "CFD study of single phase and multiphase (liquid–liquid) pump-mixer: Analyzing design parameters, flow structures and turbulence," *Chem. Eng. Sci.*, **80**, pp. 55–69, 2012.

[3] S. Huang, X. Su, J. Guo, and L. Yue, "Unsteady numerical simulation for gas-liquid two-phase flow in self-priming process of centrifugal pump," *Energy Convers. Manag.*, **85**, pp. 694–700, 2014.

[4] K. De Boer and P. A. Bahri, "Investigation of Liquid-Liquid Two Phase Flow In Biodiesel," *7th Int. Conf. on CFD in the Minerals and Process Industries CSIRO, Melbourne, Australia*, pp. 1–6, 2009.

[5] J. Karcz, M. Musia, M. Bitenc, and M. Doma, "CFD Modelling of the Fluid Flow Characteristics in an External-Loop Air-Lift Reactor," *Chemical Engineering Transactions,* **32**, pp. 1435–1440, 2013.

[6] S. Srivastava, A. Kumar, and K. Kumar, "Design analysis of Mixed Flow Pump Impeller Blades Using ANSYS and Prediction of its Parameters using Artificial Neural Network," *Procedia Eng.*, **97**, pp. 2022–2031, 2014.

[7] J. Wang, H. Zha, J. M. McDonough, and D. Zhang, "Analysis and numerical simulation of a novel gas-liquid multiphase scroll pump," *Int. J. Heat Mass Transf.*, **91**, pp. 27–36, 2015.

[8] Y. Zou, S. Ye, Y. Wang, and W. Fei, "Journal of the Taiwan Institute of Chemical Engineers CFD simulation and PIV measurement of liquid – liquid two-phase flow in pump-mix mixer," *J. Taiwan Inst. Chem. Eng.*, **60**, pp. 15–25, 2016.

[9] H. S. Alam & Bahrudin, " Integrity Evaluation of Micro- / Nanobubble Pump Impeller using Fluid-Structure Coupled Simulation," *International Journal of Engineering and Technology (IJET)*, **8**, no. 6, pp. 2594–2600, 2017..

[10] K. V. Karanth and N. Y. Sharma, "CFD Analysis of a Regenerative Pump for Performance Enhancement," *3rd World Conf. on Applied Sciences, Engineering & Technology,Kathmandu, Nepal*, pp. 3–8, 2014.

[11] T. A. Meakhail, "Numerical Study Of Unsteady Flow Characteristics In Regenerative Pump," *Journal of Engineering Sciences,* **35**, no. 4, pp. 933–943, 2007.

IOP Conf. Series: Materials Science and Engineering **204** (2017) 012008 doi:10.1088/1757-899X/204/1/012008

Production and Application of Olivine Nano-Silica in Concrete

Mardiana Oesman[1] and Haryadi[2]

[1]Civil Engineering Departmentof Politeknik Negeri Bandung, Indonesia
[2]Chemical Engineering Department of Politeknik Negeri Bandung, Indonesia

E-mail: mardianaoesman@gmail.com, haryadi@polban.ac.id

Abstract. The aim of this research was to produce nano silica by synthesis of nano silica through extraction and dissolution of ground olivine rock, and applied the nano silica in the design concrete mix. The producing process of amorphous silica used sulfuric acid as the dissolution reagent. The separation of ground olivine rock occurred when the rock was heated in a batch reactor containing sulfuric acid. The results showed that the optimum mole ratio of olivine- acid was 1: 8 wherein the weight ratio of the highest nano silica generated. The heating temperature and acid concentration influenced the mass of silica produced, that was at temperature of 90 ^{o}C and 3 M acid giving the highest yield of 44.90%. Characterization using Fourier Transform Infrared (FTIR) concluded that amorphous silica at a wavenumber of 1089 cm^{-1} indicated the presence of siloxane, Si-O-Si, stretching bond. Characterization using Scanning Electron Microscope – Energy Dispersive Spectroscopy (SEM-EDS) showed the surface and the size of the silica particles. The average size of silica particles was between 1-10 μm due to the rapid aggregation of the growing particles of nano silica into microparticles, caused of the pH control was not fully achieved.

1. Introduction

The use of amorphous micro-silica in concrete continuous to increase and, it is particularly valued in making high performance concrete. The main beneficial effects of micro-silica in concrete are increase density, reduce porosity and bleeding, and improve the bond between cement matrix and aggregates. This is due to the pozzolanic properties of micro-silica and the extremely fine particles located at close proximity to the aggregate particles [1]. Despite the beneficial properties of small silica particles in concrete, nano-silica is not yet used in common practice. Nowadays, the two most important commercial processes in the production of nano-silica are the neutralization of sodium silicate solutions with acid and the flame hydrolysis. Both processes are expensive because of the price of the raw materials and the energy requirements. Nano-silica could be applied even more widely if a new industrial, low cost, production process could be developed. However, initial research [2,4,8,9] has demonstrated that nano-silica could be produced for a low price and a sustainable method by the dissolution of olivine in acid.

Initial research [4,8,10] has revealed that nano-silica can be produced by dissolving olivine in acids. The acid is neutralized by olivine mineral, following: $(Mg, Fe)_2\ SiO_4 + 4H^+ \rightarrow Si(OH)_4 + 2(Mg, Fe)^{2+}$

The neutralization yields a mixture of a magnesium/iron salt solution, silica, unreacted olivine and inert minerals. Once the reaction is complete, the unreacted olivine and inert minerals are removed from the final suspension by sedimentation. Subsequently, the silica can be cleaned from the resulting mixture by washing and filtering. After the filtration, a cake with a 20 % solid content of nano-silica is obtained. This cake can be stored in this form, dried or redirpersed for application in concrete. The

Content from this work may be used under the terms of the Creative Commons Attribution 3.0 licence. Any further distribution of this work must maintain attribution to the author(s) and the title of the work, journal citation and DOI.
Published under licence by IOP Publishing Ltd

nano-silica produced has a specific surface area between 100 and 300 m^2/g and a particle size between 10 and 25 nm. These particles are agglomerated in clusters forming a mesoporous material with an average pore diameter around 20 nm. The impurity content is below 5%, and the silica yield of the process in the range of 54–83%. When the separation of the silica is carried out by filtration, the textural properties of the material have a direct influence on the separation efficiency.The texture and specific surface area of nano-silica depends on several factors of the process, the main ones being the kinetics of the dissolution of olivine and the washing steps of the nano-silica. Thus, the properties of this nano-silica can be tailored by changing the process conditions [3].
Porro*et al.* [7] investigated the effect of the size of SiO_2 nanoparticles used at different dosages on the performance of cement pastes. It was demonstrated that the compressive strength of the cement pastes increased with the reduction of the particle size. This improvement was attributed to the formation of larger C-S-H silicate chains in mixtures with nano-SiO_2, as seen in figure 5. Nano silica is one of the blending materials used to improve the properties of concrete. Nanosilica concrete is stickier than normal concrete due to large surface area of nanosilica, also it has better permeability resistance.

2. Methodology

This research was aimed to obtain a specific amorphous silica in the rock olivine (Mg, Fe) $_2SiO_4$).The experiment comprised of two phases. The first phase was to make amorphous nano silica from the ground olivine rock using sulfuric acid as the reagent dissolution fluid according to Lazaro [3]. The ground olivine rock was supplied from Pomalaa region, South Sulawesi. Experiments were carried out at the following temperatures: 70, 80 and 90 °C.The olivine rock smoothed with 0.2 mm sieve passes and was used of 10 gram on each experiment for various temperatures. The optimum point of olivine-sulfuric acid ratio must first be determined at which the olivine-acid mole ratio of how much the maximum weight of silica is gained. The yield was obtained by varying the reaction temperature and the concentration of acid used for the reaction. Acidity was controlled, because the pH should be in the range of 0.5 to 1.0. After the heating process for 4 hours, resulting slurry of silica minerals. Then the silica mud could be washed into slurry of silica. Once the minerals dissolved in the acid have been removed, then decantation process was done. This solution was then taken to be filtered using Whatman 42. The washing and filtering processes were repeated four times until the washing pH was in the range of 6 to 7. Then, silica mud was dried in an oven at 110°C for 2 hours. Silica was sized between the micrometer to nanometer range that was characterized using SEM-EDS to observe the size and the surface microstructures. FTIR instrument was used to characterize the functional groups of silica particles produced.
The second phase was compressive strength tests of concrete specimens. The nanoolivin silica was applied in concrete mix design as cement substitution. There were two types of silica size those were 700 nm - 950 nm (type I), and 1000 nm - 10,000 nm (Type II). The specimens were cubic with 150 mm side length. Besides nano-silica, fly ashwas also used as a substitute for cement. Mix concrete design based on ACI 211.1-97 with the value of slump was 50mm, and water-cement ratio was 0.3. The compression strength was design at 40 MPain 28 days.The maximum coarse aggregate sizewas 13 mm. The concrete specimens without nano silica were constructed to be used as a reference to compare the behavior of the specimens.The compressive strength testing performed at the age of 7 days and 28 days using UTM.

Table 1. Optimization of molar

Molar ratio of Olivine: Acid	Yield of Silica, (%)
1 : 4	16.90
1 : 5	27,90
1 : 6	25,00
1 : 7	29,20
1 : 8	44,50
1 : 10	39,90
1 : 12	31,20

Table 2. Optimization of temperature

Operation Temperature, °C	Yield of Silica, (%)
70	30,60
80	44,60
90	44,90

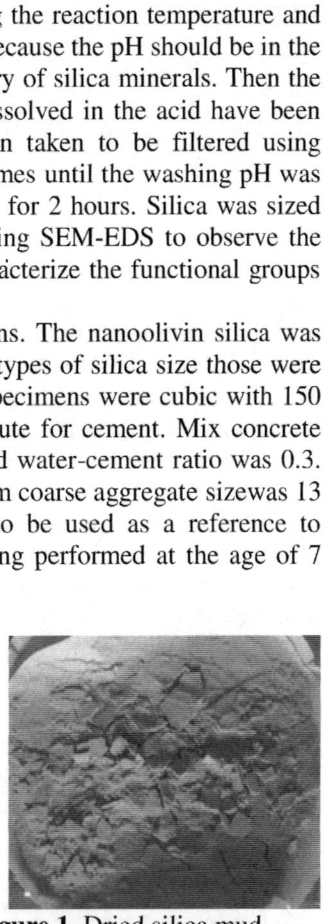

Figure 1. Dried silica mud

3. Result and Discussion

3.1 Process of nano-silica production

In this research there were two steps to get optimum condition for silica production from ground Olivine rocks followed by dissolution namely molar ratio of Olivine over acid concentration and operation temperature. The series of molar ratio for Olivine to concentration of sulfuric acid was conducted to obtain the optimum condition for silica production using 3M sulfuric acid at 80°C and the results are listed in Table 1. It is indicated that yield of silica tends to increase in accordance with the increasing of sulfuric acid quantity. The maximum weight of silica produced was in condition: 1 mole olivine reacts with 8 moles of sulfuric acid at temperature of 80°C producing 44.50 % yield of silica. For operation temperature variation, it shows that the maximum operation temperature for dissolution was 90°C which is close to the finding of Lazaro, et.al [3], as shown in Table 2. In addition, the result of silica separation process from the ground olivine rock was dried mud as shown in figure 4. The increasing temperature and sulfuric acid concentration results in an increase in the mass and surface area of nanosilica [6]. Separation bonding silica from olivine is controlled by surface reaction and is dependent on the surface area of olivine, the hydrogen ion activity, and temperature. Thus, increasing the surface area of olivine, the hydrogen ion activity and the reaction temperature will increase the rate of reaction [2].

3.2 Characterization of Nanosilica by FTIR

Fourier Transform Infrared (FTIR) investigation aims to determine the characteristics of nanosilica by observing the functional groups in the silica sample. The infrared spectrum of silica produced from Olivine can be seen in figure 2. The main peak at 3446.7 cm^{-1} is observed which is assigned the silanol groups (Si-OH) and adsorbed water. A sharp peak at 1639 cm^{-1} assigned for –C-O stretching vibration caused due to the atmospheric CO_2 adsorbed by the amorphous nanosilica. A strong band at 1089 cm^{-1} is caused by the presence of siloxane (Si-O-Si) stretching that suggest the local bonding structure of Si and O atoms at nanoparticles size. A peak corresponding to silanol groups (Si-OH) stretching is observed at 947 cm^{-1}. A sharp and intense peak at 796 cm^{-1} is caused by the presence of Si-O bending vibrations. The peak at 461 cm^{-1} is assigned by the Si-O out plane bending vibrations. The functional groups indicate specific functional groups contained in nano-silica. The silica separation from olivine rock has been successfully done by observing at the FTIR spectrum. The FTIR results can be summarized that the silica resulting from the extraction process of olivine rock has the functional group Si-OH (silanol), Si-O-Si (siloxane) and the OH group (hydroxyl) with the presence of adsorbed water in silicaparticle.

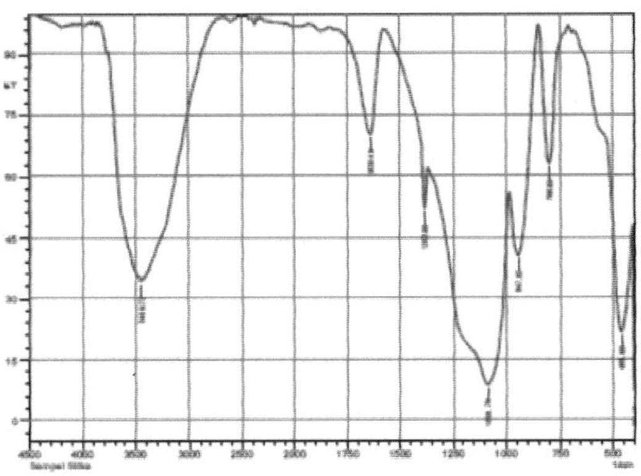

Figure 2. The analysis results of functional groups with FTIR ratio

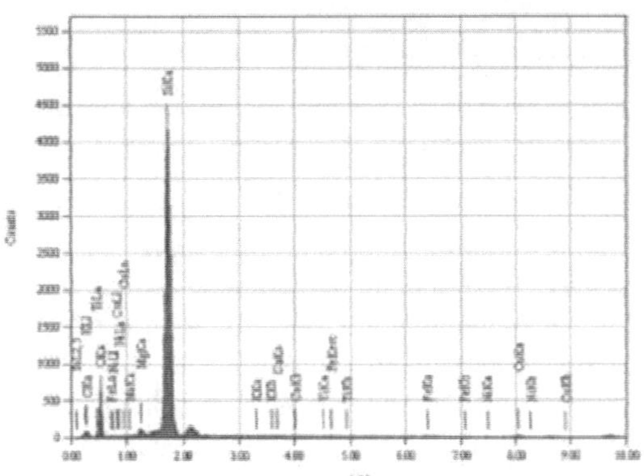

Figue 3. EDS spectrum of silica product of functional groups with FTIR ratio

3.3 Characterization of silica by SEM-EDS

The mass percent of eachdried compounds in a silica sample characterized using SEM-EDS. SEM-EDS analysis performed to obtain the surface morphology of the sample with a very high resolution. Figure 3 shows EDS spectrum with a variety of elements and compounds in the silica sample. It shows that the compounds are dominated by SiO_2 for about 82.3 % followed by C, CuO, MgO of 11.9, 2.9 and 1.5 % by mass respectively. Figure4 shows the morphology of silica particles extracted from olivine rocks after dissolution process observed bySEM-EDS with a magnification of 500 times. The figure shows the scale of the large chunk silica size is 500 μm. One large chunk is a collection of several small-sized silica particles are fused. Actually, expected amorphous silica obtained with an average size in nanometers. This occured due to the nature of the silica particles that relatedto Ostwald ripening, in which the larger particles grow at the expense of smaller particles [2]. Particles are first formed at nanometer-sized and form a ball up micro-sized particles, through the mechanism of aggregation. This should be addressed during the reaction process; the acid concentration must be maintained at concentrations below 0.1 mol / L by controlling pH through addition of $Na_2B_4O_7$ at regular intervals [2]. Figure 3 also displays the silica with a magnification of 300 times viewable size of 10 μm silica indicating various silica particle sizein the range of 100-1000 nm.

3.4 *Compressive test of nano-silica concrete.*

The results of compressive strength were conducted at ages of 7 days and 28 days. Table 3 shows the strengths of concrete with silica (type I and II) at the age of 7 days were 60% strength at 28 days.It is consistent with the results of research in general that is approximately 60-70% of concrete strength 28 days. The table indicates that the highest compressive strength achieved by the specimen without silica. This is not consistent with the predictions expected in this study, which is to achieve a high compressive strength concrete. This happened because the conditions predicted silica size obtained from the process of making nanosilica has not been reached properly.

Figure 4. SEM image of nanosilica extracted from Olivine rock

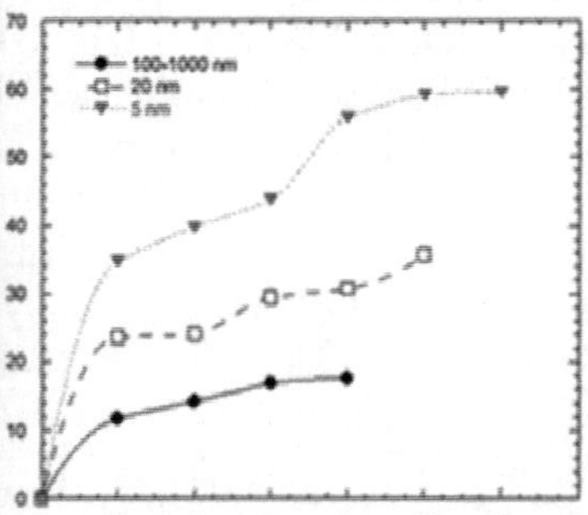

Figure 5. The effect of the nanosilica size to f'c [7]

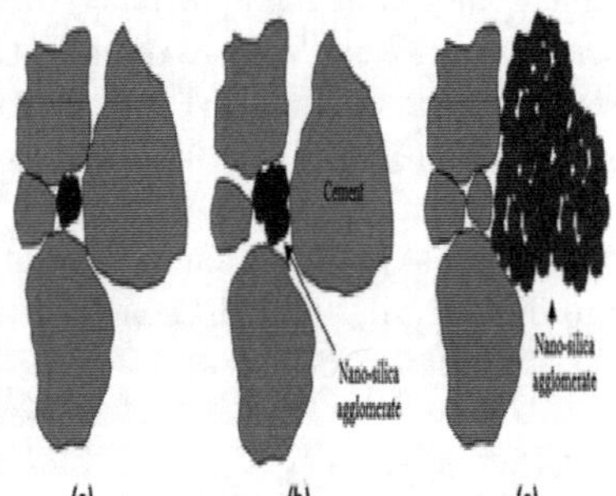

Figure 6. Illustration of the filling effect of the nano-silica agglomerates [5]

Porro et. al.[7] said that compressive strength of concrete with silica as subtitution of cement can be increase if the silica size is under 100nm, as shown in figure 5. Actually, the product of silica size from the first phase of this research was in the rangeof 100-1000 nm so that the compressive strength of the concrete in this research was less than the concrete without silica. Moreover, Kong et al. [5] described the effect of particle size on the nano concrete, namely (a) small clumps of nano particles behave as a filler which remove the water trapped in the fresh pasta, (b) larger nanoparticles cannot behave as a filler, which led to an increase in cavities, and (c) large nanoparticles can absorb water which initially may affect viscosity paste. From these statements, it can be predicted how the nano silica condition that occurs in this study. Nanosilica condition can agglomerate into micro and even macro size, so that they cannot behave as a filler as expected. Thus, the silica can not decrease the porosity happened, instead to an additional porosity formed. This causes the concrete has a high porosity; density occurring impact is low, and the low concrete strength (figure 6). Therefore, the study of silica nanoscale manufacturing process must be continued; so it can be applied to the manufacture of nano silica concrete to achieve a high performance concrete.

Tabel 3: Compressive strength

Specimens	7 days	28 days
Type I (nano)	12.57 MPa	22.93 MPa
Type II (micro)	16.00 MPa	25.53 MPa
Without Silica	24.78 MPa	34.97 MPa

Conclusion

The mole ratio of olivine-optimum acid is 1: 8 wherein the weight ratio of the highest nanosilica generated. Effect of heating temperature and acid concentration influence the weight of silica produced where at 90°C and acid produces 3 Molar yieldsthe highest of 44.90%. Characterization using FTIR concluded that amorphous silica at wavenumber of 1089 cm^{-1} corresponding to stretching vibration of siloxane bond, Si-O-Si. Characterization using SEM-EDS displays the average size of the silica particles is 1-10 μ m due to the rapid aggregation of the growing particles of nanosilica into microparticles where pH control is not fully achieved. The average size of silica particles in the range 0.7-10 μm has not reached the optimum of nanosilica concrete.Therefore, it needs to conduct more research toget silica particle size below 0.1 μm forpreparing high performanced concrete.

References

[1] A.M. Neville, 1996, Properties of concrete, Fourth Edition, John Wiley & Sons.

[2] A.Lazaro, H.J.H, 2011, Brouwers,A new method for the production of colloidal nano-silica, Colloids and Materials, *The 1st symposium on Colloids and Materials*, Amsterdam.

[3] A. Lazaro, H.J.H. Brouwers, G. Quercia, J.W. Geus, The properties of amorphous nano-silica synthesized by the dissolution of olivin, *Chemical Engineering Journal* 211-212, 2012, 112–121.

[4] D.J. Lieftink, 1997,The preparation and characterization of silica from acid treatment of olivine, PhD thesis, Utrecht University.

[5] D. Kong, Y.Su, X.Du, Y.Yang, S.Wei, S.P. Shah, Influence of nano-silica agglomeration on fresh properties of cement pastes, *Construction and Building Materials*, **43**, 2013, 557–562.

[6] L.P. Singh, A. Goelb, S.K. Bhattacharyya,G. Mishraa, 2015, Improving the durability of cementitious materials using silica nanoparticle. *Nanotechnology in Construction*.

[7] Porro, A., Dolado, J.S., Campillo, I., Erkizia, E., de Miguel, Y., Sáez de Ibarra, Y., Ayuela, A. 2005, Effects of nanosilica additions on cement pastes: Applications of nanotechnology in concrete design, Thomas Telford, London.

[8] R.C.L. Jonckbloedt, 1997, The dissolution of olivine in acid, a cost effective process for the elimination of waste acids, PhD thesis, Utrecht University.

[9] R.D. Schuiling, 1986, A method for neutralizing waste sulphuric acid by adding a silicate", US 1987/4707348.

[10] Van Herk, J., Pietersen H.S., Schuiling R.D., 1989, Neutralization of industrial waste acids with olivine — The dissolution of forsteritic olivine at 40–70°C", *Chemical Geology*, **76**: p 341.

[11] Gorrepati, E., Wongthahan, P., Raha.S., and Fogler, H S., Siliica Precipitation in Acidic Solutions: Mechanism, pH Effect, and Salt Effect, Langmuir, **26 (13)**, 2010, pp 10467-10474.

IOP Conf. Series: Materials Science and Engineering **204** (2017) 012009 doi:10.1088/1757-899X/204/1/012009

Synthesis, Characterization and Visible Light Photocatalytic of Fe_3O_4/CuO/TiO_2/Ag Nanocomposites

Malleo Fauzian[1, 2], Ardiansyah Taufik[1,2] and Rosari Saleh[1,2,]

[1]Departemen Fisika, Fakultas MIPA-Universitas Indonesia, 16424 Depok, Indonesia
[2]Integrated Laboratory of Energy and Environment, Fakultas MIPA-Universitas Indonesia, 16424 Depok, Indonesia

E-mail: malleo.e604@gmail.com, ardiansyah.taufik@ui.ac.id, rosari.saleh@gmail.com

Abstract. The sol-gel method was used to synthesize Fe_3O_4/CuO/TiO_2/Ag visible light photocatalyst. X-ray diffraction (XRD) and ultraviolet–visible (UV–Vis) spectroscopy were used to characterize the photocatalyst. From the UV–Vis absorption spectra, we found the surface plasmon resonance (SPR) to be around ~440 nm. The results indicate that the nanocomposites contained a combination of the desired nanocomposites. Their photocatalytic activity was evaluated from the degradation of methylene blue. The influence on the activity of the Fe_3O_4/CuO/TiO_2/Ag nanocomposites, such as weight percentage of Ag, was also studied. Fe_3O_4/CuO/TiO_2/Ag with 25wt% of Ag exhibited the best photocatalytic activity under visible light irradiation. The experiment results showed that the main active species in the photodegradation mechanism in methylene blue is holes. Finally, the nanocomposites exhibited good stability after being reused four times in a cycling process.

1. Introduction

With the recent rise in environmental problems, the treatment of various organic pollutants in wastewater is a major issue. Semiconductors have recently been used for degradation of organic pollutants in wastewater. Among all semiconductors, TiO_2 has been widely use in the photodegradation of organic pollutants because of its nontoxicity, low cost, and good photocatalytic activity [1-4]. However, TiO_2 can only be excited under ultraviolet irradiation, resulting in low utilization of solar energy, because of which TiO_2 has a high electron–hole recombination rate. This is a disadvantage in the utilization of the photocatalytic activity of TiO_2 [5-8]. This limitation can be overcome by employing methods such as doping [9], coupling with another semiconductor [10], and deposition of noble metals [11]. Our research group has reported the synthesis and characterization of coupled TiO_2 with another semiconductor, CuO and Fe_3O_4, creating Fe_3O_4/CuO/TiO_2nanocomposites with good photocatalytic activity under ultraviolet irradiation. However, the degradation efficiency in the visible light region still requires improvement. Noble metals, such as Au, Ag, and Pt, have often been coupled with TiO_2 nanoparticles to enhance the photocatalytic efficiency under visible light irradiation [12, 13]. The noble metals can promote the interfacial charge transfer process in the composites system and electron traps.

The research aims to study the synthesis and characterization of Fe_3O_4/CuO/TiO_2/Ag nanocomposites and to investigate the effect of incorporating Ag in Fe_3O_4/CuO/TiO_2 as a catalyst for degradation of methylene blue under visible light irradiation. The influence of different weight percentages of Ag in Fe_3O_4/CuO/TiO_2 on the photocatalytic efficiency was also evaluated. Radical scavengers were employed to provide an insight into the photodegradation mechanism through different active species.

Content from this work may be used under the terms of the Creative Commons Attribution 3.0 licence. Any further distribution of this work must maintain attribution to the author(s) and the title of the work, journal citation and DOI.
Published under licence by IOP Publishing Ltd

IOP Conf. Series: Materials Science and Engineering **204** (2017) 012009 doi:10.1088/1757-899X/204/1/012009

Finally, also investigated the stability of nanocomposites in photocatalytic process, and the catalyst was reused four times in the cycling processes.

2. Experimental

In this research,materials were used such as iron (II) sulfate heptahydrate ($FeSO_4$•$7H_2O$, 99%), copper sulfate pentahydrate ($CuSO_4$•$5H_2O$, 99%), titanium dioxide (TiO_2), silver nitrate ($AgNO_3$), sodium hydroxide (NaOH), ethanol, and ethylene glycol (EG) purchased from Merck. All chemical reagents (Merck) were of analytical grade and used without further purification. $Fe_3O_4/CuO/TiO_2/Ag$ nanocomposites were prepared using the sol-gel method [14] with further modification. First, TiO_2 and NaOH were mixed with distilled water. Further, the solution was stirred at 80˚C using a magnetic stirrer. Fe_3O_4, CuO, and Ag were dispersed into a solution of ethanol and distilled water and added into the suspension at a particular ratio. After 2 h of stirring and heating, the mixed solution was collected, washed, aged, and then dried at 125 °C for 1 h under vacuum. In this experiment, the molar ratio of Fe_3O_4, CuO, and TiO_2 was 0.5:1:1, and the amount of Ag in $Fe_3O_4/CuO/TiO_2$ was varied from 5wt% to 25wt% – FCT/Ag 5wt%, FCT/Ag 15wt%, and FCT/Ag 25wt%.

To determine the structure and phase of the samples, X-ray diffraction (XRD) measurements were carried out using a RigakuMiniflex 600 X-ray diffractometer with monochromatic Cu Kα radiation, and the optical properties of the samples were studied using UV-Vis spectroscopy (Hitachi UH-5300) in the 200–750 nm range.

The photocatalytic activity was evaluated from the degradation of the organic pollutant, which in this research is methylene blue (MB), in the presence $Fe_3O_4/CuO/TiO_2/Ag$ nanocomposites in an aqueous solution under visible light using a 50 W halogen lamp. In the experiment, 40 mg of $Fe_3O_4/CuO/TiO_2/Ag$ nanocomposites was added to 100 ml of MB with an initial concentration 20 mg/L. Then, the mixture was stirred for 30 min in the dark to establish desorption–adsorption equilibrium between solution and photocatalyst. The lamp was then turned on, and this was considered 'time zero" for the reaction. In this research, the photocatalytic reaction proceeded for 2 h, and samples were collected at regular intervals. To examine the stability of the samples, they were reused four times. After testing, the sample was separated with a solution using an external magnet. A decrease in the concentration of the MB solution was measured by a Shimadzu spectrophotometer at the wavelength of 663 nm.

3. Results and discussion

X-ray diffraction spectra of $Fe_3O_4/CuO/TiO_2/Ag$ nanocomposites with various weight percentages of Ag are plotted in Fig 1a. The diffraction peaks at $2\theta = 25.24°$ and 37.8°, which are attributed to the (100) and (004) planes, respectively, are the crystal faces of anatase TiO_2. Phase cubic spinel Fe_3O_4 was detected at ($2\theta = 30.14°$, 35.49°, 43.28°, 53.76°, 57.20°, and 62.83°) and monoclinic CuO at ($2\theta =$ 38.81° and 48.7°). After addition of Ag 25wt%, new peaks were detected at $2\theta = 37.52°$, 44.61°, and 64.57°, which represents the cubic structure of Ag. The absence of Ag peaks in other nanocomposites is because the amount of Ag is too low compared with the other materials. Table 1 presents the crystallite size calculated according to the Scherrer equation [15-16] and the unit cell parameters of all samples. As can be seen, in the sample with 5–15 wt% of Ag, the crystallite size of Ag is too small to be detected.

The optical properties of the as-prepared samples were determined from the UV–vis absorbance spectra. As shown in Fig 1b, $Fe_3O_4/CuO/TiO_2/Ag$ nanocomposites with various weight percentages of Ag have absorption peaks at ~440 nm, which display typical surface plasmon resonance (SPR) [17]. However, it indicates the presence of Ag in the nanocomposites. Coupling Ag with a composite is expected to create a Schottky junction. Schottky junctions mainly facilitate electron–hole separation and promote interfacial electron transfer process. Furthermore, in nanocomposites made of Ag and TiO_2, given that Ag is a plasmonic system, one could expect improved absorption, primarily in the visible region. This phenomenon may be beneficial for photocatalytic activity in the visible region.

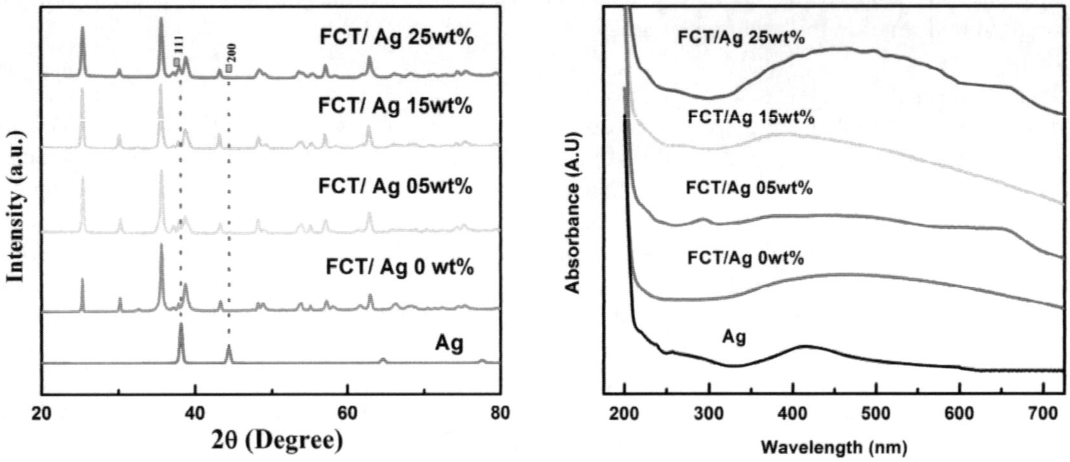

(a) (b)

Figure 1. (a) XRD spectra and **(b)** Uv-Vis absorbance spectra of $Fe_3O_4/CuO/TiO_2/Ag$ with various wight percent of Ag.

Table 1. Grain Size of $Fe_3O_4/CuO/TiO_2/Ag$ variation weight percent

Samples	Grain Size (nm)			
	Fe_3O_4	CuO	TiO_2	Ag
$Fe_3O_4/CuO/TiO_2$	42	15	79	-
$Fe_3O_4/CuO/TiO_2/Ag$ 5wt%	40	15	79	-
$Fe_3O_4/CuO/TiO_2/Ag$ 15wt%	40	15	78	-
$Fe_3O_4/CuO/TiO_2/Ag$ 25wt%	40	14	75	13

Fig 2a shows the photocatalytic activity of $Fe_3O_4/CuO/TiO_2/Ag$ nanocomposites with various weight percentages of Ag based on the degradation of MB under visible irradiation. As can be seen, $Fe_3O_4/CuO/TiO_2/Ag$ nanocomposites with 25 wt% of Ag have the best photocatlytic activity under visible irradiation. It can degrade up to 100% of MB. Incorporation of Ag in the nanocomposites improves the photocatalytic activity owing to SPR in the nanocomposites [18, 19], and with the addition of Ag in the nanocomposites, there is an improvement in the visible light absorption of the nanocomposites.

The influence of dosage catalyst concentration on the degradation of MB using $Fe_3O_4/CuO/TiO_2/Ag$ nanocomposites with 25 wt% of Ag in terms of the photocatalytic activity under visible light irradiation is shown in Fig 2b. The dosage catalyst concentration were varied from 0.1 to 0.4 g/L. With the increase in the dosage catalyst, the degradation of MB was better than that in the case of photocatalytic degradation. A dosage catalyst of 0.4 g/L enhanced the degradation efficiency of MB to 100% within 2 h. This phenomenon may be due to the increase in the surface area of the catalyst to adsorb the MB.

To examine the stability of the catalyst in photocatalytic activity under visible light irradiation, the catalyst was reused four times in a cycling process under the optimum conditions. In this research, $Fe_3O_4/CuO/TiO_2/Ag$ with 25 wt% of Ag could be easily separated from the solution using a permanent magnetic bar. External magnetism is an easy and efficient way to separate the catalyst from the suspension system. The catalyst was reused with the same quantity of fresh MB after each run.

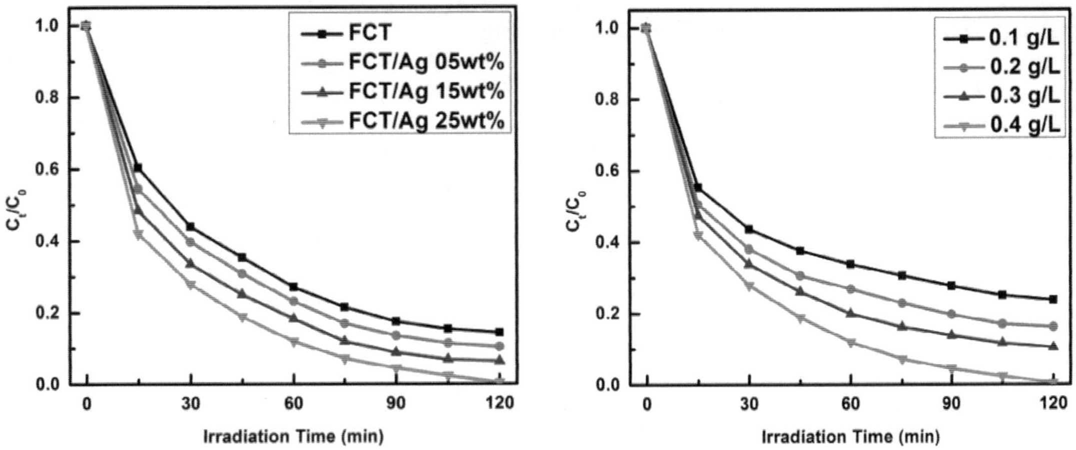

(a) **(b)**

Figure 2. **(a)** Photocatalytic activity and Influence dosage catalyst concentration of $Fe_3O_4/CuO/TiO_2/Ag$ 25wt% to degradation of MB under visible light irradiation.

The results are shown in Fig 3a. As can be seen, after the fourth cycle, degradation of MB slightly declines, only by ~10%. So, it can concluded that the nanocomposites are stable material after reused fourth time cycling process because contained Fe_3O_4 in a sample that has a value of magnetization is pretty good, so in the process of separating the sample with a solution becomes easier and the sample is not a lot of wasted.

It is importance to detect the main active oxidative species in the degradation of MB to determine the mechanism of the catalytic activity. Therefore, we control the experiment with the addition of scavengers for electrons, holes, and hydroxyl radicals into the solution under the optimum conditions. Fig 3b shows the degradation of MB in presence of tert-butyl alcohol, diammonium oxalate, and $Na_2S_2O_8$ scavenger for hydroxyl radicals, holes, and electrons, respectively. As can be seen in the figure, after addition of diammonium oxalate in the MB solution, the photodegradation of MB decreased. Thus, the main active species in the experiment is holes.

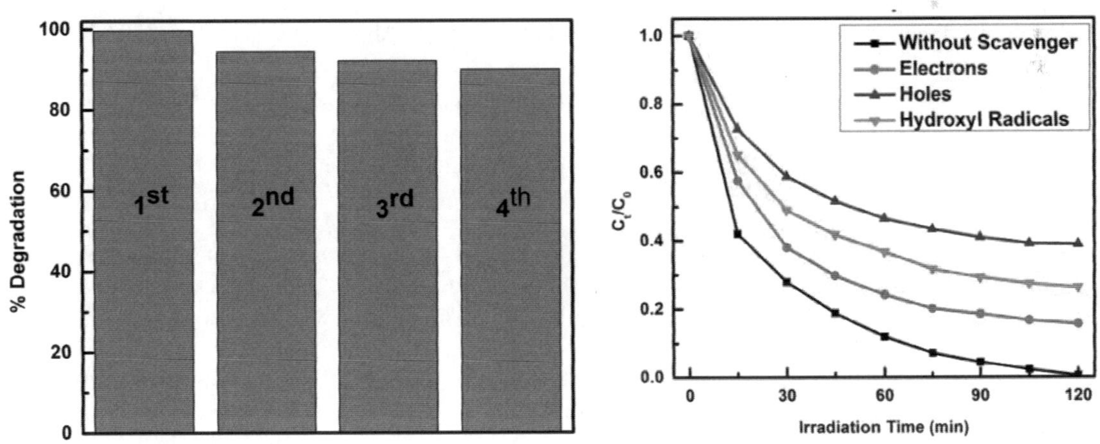

(a) **(b)**

Figure 3. **(a)** Reusability and **(b)** Effect of different Scavenger of $Fe_3O_4/CuO/TiO_2/Ag$ 25wt% under visible light irradiation.

4. Conclusion

$Fe_3O_4/CuO/TiO_2/Ag$ nanocomposites with various weight percent of Ag were successfully synthesized using sol gel method and active catalyst under visible irradiation for degradation of Methylene Blue (MB). The sample were characterized by using X-ray diffraction (XRD) and Uv-Vis absorbance. It confirmed that the final $Fe_3O_4/CuO/TiO_2/Ag$ nanocomposites with various weight percent of Ag. $Fe_3O_4/CuO/TiO_2/Ag$ nanocomposites with 25 weight percent of Ag the best photocatalytic activity under visible light to 100% within 2h, this happens due to Surface Plasmons Resonance in nanocomposites and nanocomposites can absorb visible light is much better than the other nanocomposites. Increase dosage catalyst to 0.4 g/L the degradation efficiency of MB because to increase surface area of catalyst to adsorb the MB. Holes is the main active oxidative species in photocatalytic activity on degradation of MB. The catalyst has good stability after reused fourth cycled degradation of MB slightly decline only ~10%.

References

[1] M.D. Hernández-Alonso, F. Fresno, S. Suárez, J.M. Coronado, *Energy Environ.* Sci. **2** (2009) pp 1231–1257.
[2] E. Balaur, J.M. Macak, H. Tsuchiya, P. Schmukiet, J. *Mater. Chem.* **15** (2005) pp 4488–4491.
[3] D.V. Bavykin, V.N. Parmon, A.A. Lapkin, , J. *Mater. Chem.* **14** (2004) pp 3370–3377.
[4] Z. Liu, D.D. Sun, P. Guo, J.O. Leckie, *Nano. Lett.* **7** (2007) pp 1081–1085.
[5] B. Ohtani, R.M. Bowman, D.P. Colombo, H. Kominami, H. Noguchi, K. Uosaki, *Chem. Lett.* **27** (1998) pp 579–580.
[6] S. Ikeda, N. Sugiyama, B. Pal, G. Marci, L. Palmisano, H. Noguchi, *Phys. Chem. Chem. Phys.* **3** (2001) pp 267–273.
[7] A. Fujishima, X.T. Zhang, D.A. Tryk, *Surf. Sci. Rep.* **63** (2008) pp 515–582
[8] J. Li, W, H. Ma, Y.P. Huang, X. Tao, J.C. Zhao, Y.M. Xu, *Appl. Catal. B: Environ.* **48** (2004) pp 17-24
[9] T. Saidani, M. Zaabat, M.S. Aida, B. Boudine, *Superlattices and Microstructures* **88** (2015) pp 315-322
[10] T. Xu, L. Zhang, H. Cheng, Y. Zhu, *Applied Catalysis B: Environmental* **101** (2011) pp 382 -387
[11] S. Mun Lam, J. Chung Sin, A. ZuhairiAbudllah, A. Rahman Mohamed, *Separataion and Purification Technology* **132** (2014) pp 378-387
[12] L. Kumaresan, M. Mahalakshmi, M. Palanichamy, V. Synthesis, *Ind. Eng. Chem. Res.* **49** (2010) pp 1480–1485.
[13] M.R. Mohammadizadeh, M. Bagheri, S. Aghabagheri, Y. Abdi, *Appl. Surf. Sci.* **350** (2015) pp 43–49.
[14] S. A Jalaludin, S. A Arifin, R. Saleh, *Material Science Forum* **827** (2015) pp 13-18
[15] R.Y. Hong, S.Z. Zhang, G.Q. Di, H.Z. Li, Y. Zheng, J. Ding, D, G. Wei, Preparation, *Materials Research Bulletin* **43** (2008) pp 2457-2468.
[16] A. Hamrouni, H. Lachheb, A. Houas, *Materials Science and Engineering* B **178** (2013) pp 1371- 1379.
[17] S. S. Patil, M. G. Mali, M. S. Tamboli, D. R. Patil, M. V. Kulkarni, H. Yoon, H. Kim, S. S. Al-Deyab, S. S. Yoon, S. S. Kolekar, B. B. Kale, *Catalysis Today* **260** (2016) pp 126-134
[18] X. Zhang, L. Wang, C. Liu, Y. Ding, S. Zhang, Y. Zeng, Y. Liu, S. Luo, *Journal of Hazardous Materials* **313** (2016) pp 244-252
[19] J. Chen, H. Che, K. Huang, C. Liu, W. Shi, *Applied Catalysis B: Enviromental* **192** (2016) pp 134-144

IOP Conf. Series: Materials Science and Engineering **204** (2017) 012010 doi:10.1088/1757-899X/204/1/012010

Influence of surface activated carbon nano fibres on mechanical properties of poly ether ketone (PEK)

G Ajeesh[1], S Bhowmik[1], V Sivakumar[1] andL Varshney[2]

[1] Department of Aerospace Engineering, Amrita University, Coimbatore, 641112, India
[2]Radiation Technology Development Division, Bhabha Atomic Research Centre, Mumbai, 400085, India

Email : g_ajeesh@cb.amrita.edu,
b_shantanu@cb.amrita.edu,v_sivakumar@cb.amrita.edu, lalitv@barc.gov.in

Abstract. This investigation highlights different surface functionalization processes of Carbon Nano Fibres (CNF's) and their effects on mechanical properties of Polyetherketone (PEK) nano composite. Surfaces of CNF's were modified by low pressure plasma process. There is a significant change in physico-chemical characteristics of CNF's after low plasma treatment as evident from Transmission Electron Microscopy (TEM) and Fourier Transform infrared Spectroscopic (FTIR)studies. Significant modification in surface morphology and oxygen functionalities are observed as a result of surface modification. There is a significant increase in mechanical properties of high performance polymeric nano composites when surface functionalized CNF's are dispersed in polymeric matrix.

1. Introduction

Polymer matrix composites (PMC's) are a judicious combination of matrix and reinforcement. The resistance of the PMC towards most of degrading environment is determined by the matrix. Few such processes include impact damage, delamination, water absorption, chemical attack, and high-temperature creep. Thus, the matrix is typically the weak link in the PMC structure. The reinforcement is responsible for providing high strength to the composites. The most common reinforcements presently in use are glass, graphite and aramid. In the last decade, extensive researches were carried out on the use of nano particles as reinforcements.

The major issue pertaining to the use of nano particles is the interfacial adhesion and dispersion of the nano particles within the polymers. The interphase is the region between the polymer and reinforcement responsible for load transfer [1]. These problems can often be controlled by using an appropriate coating on the reinforcing fibers.

This investigation highlights the effect of plasma modification of carbon nano fibres on the various properties of PEK high performance composites.

2. Experimental

In this investigation, two types of composites were prepared. PEK with unmodified CNF's and PEK with plasma treated CNF's subjected to mechanical testing. CNF's were dispersed in PEK matrix with 2% weight ratio. The matrix and reinforcement were compounded in a twin screw extruder at a

Content from this work may be used under the terms of the Creative Commons Attribution 3.0 licence. Any further distribution of this work must maintain attribution to the author(s) and the title of the work, journal citation and DOI.
Published under licence by IOP Publishing Ltd

IOP Conf. Series: Materials Science and Engineering **204** (2017) 012010 doi:10.1088/1757-899X/204/1/012010

temperature of 400 °C. The compounded material was fed into a 1500Bar injection molding machine. Two kinds of samples were developed from injection molding machine.

FTIR spectra were recorded by an apparatus developed by Nicolet iS10 spectrometer. FTIR was performed to find the presence of new functional groups due to the exposure to plasma. Transmission spectra in the range of 4000–400 cm^{-1} were studied with a resolution of 4 cm^{-1}.

Transmission electron microscopy was to study the dispersion of modified and non modified CNF's in the polymeric composite samples.

3. Results and Discussion

3.1 FTIR results

Fourier Transform Infrared Spectroscopy (FTIR) is a method to obtain an infrared spectrum of absorption or emittance of solid or liquid. The FTIR spectrum reveals significant information on the formation of new functional groups in CNF's as a result of the interaction with plasma. The FTIR images of unmodified and modified CNF's is as shown in Figure 1 and 2.

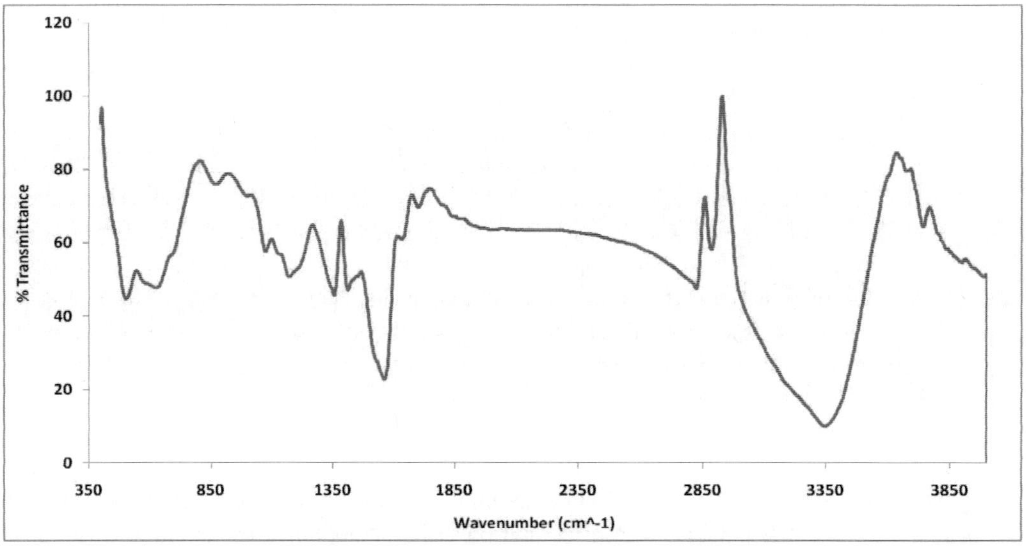

Figure 1. FTIR plot for unmodified CNF.

Figure 2.FTIR plot for modified CNF.

IOP Conf. Series: Materials Science and Engineering **204** (2017) 012010 doi:10.1088/1757-899X/204/1/012010

The above FTIR figure shows that there is a significant change in the oxygen functional groups when exposed to oxygen plasma. This clearly indicates that the surface of CNF is being modified.Therefore, by functionalizing the nano particles, the dispersion and interfacial bonding between CNF's and polymers can be enhanced [10]. Many studies have been reported using Single Wall Carbon Nanotubes and Multi Wall Carbon Nanotubes as reinforcements for polymeric composite materials to improve the properties of the material [2-4].

3.2 Tensile strength

The polymeric composites reinforced with unmodified and modified MWCNT's were subjected to tensile test. The specimens were tested on a ZWICK ROELL machine with a capacity of 10KN. The results of composites with unmodified and modified CNF are as shown in Figure 3.

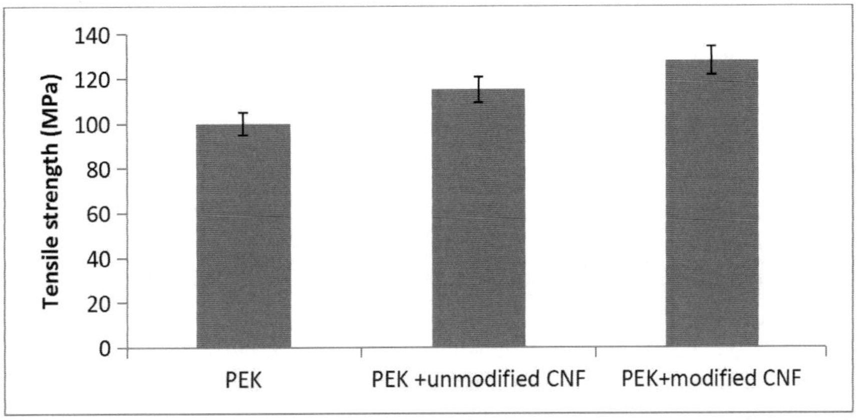

Figure 3. Mechanical strength of composites.

Figure 3 clearly indicates that there is a significant increase in the mechanical strength of composites on reinforcing modified CNF. This increase in the mechanical strength is attributed to the high interfacial adhesion developed between polymer and reinforcement. It is also observed that there is an increase in the tensile strength of PEK composites due to the influence of carbon nano fibre. CNF when reinforced in PEK gives increased strength due to the exceptional modulus of the nano filler combined with the large surface area [5-6]. In this context, research on PEK reveals that the backbone of PEK is especially rigid because of dominant phenyl groups. The side groups and the regularity of the polymer are basically semi crystalline in nature. This semi crystalline structure of PEK, coupled with the presence of dominating phenyl functional groups in the polymeric chain, provides PEK with exceptional resistance to radiation similar to that of PEEK. The energy of the radiation is thus dissipated within the crystallites through a phenomenon known as the "cage effect" and within the many aromatic structures of the phenyl groups, thus avoiding damaging the structure and the bonds of the molecules [7-8]. Therefore, it can be concluded that the presence of a strong and rigid backbone may explain why the mechanical properties are not affected after exposure to the aggressive environments [9].

3.3 TEM analysis

TEM is performed on the composite samples to study the effect of modification of nano particles in the dispersion of nano particles in composites. Figure 4 and Figure 5 shows the TEM micrographs of polymeric sample.

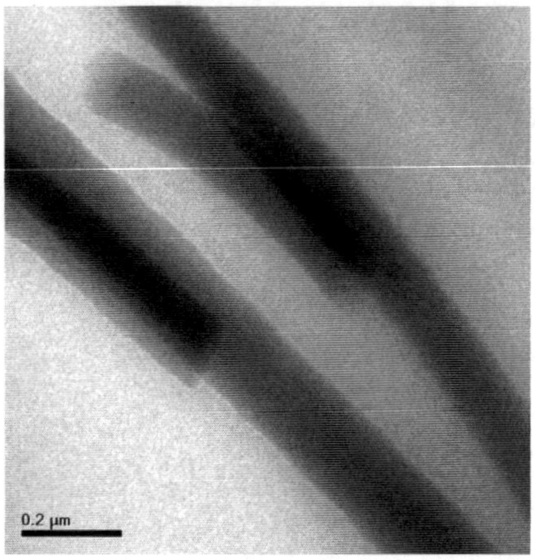

Figure 4.TEM of unmodified CNF composite

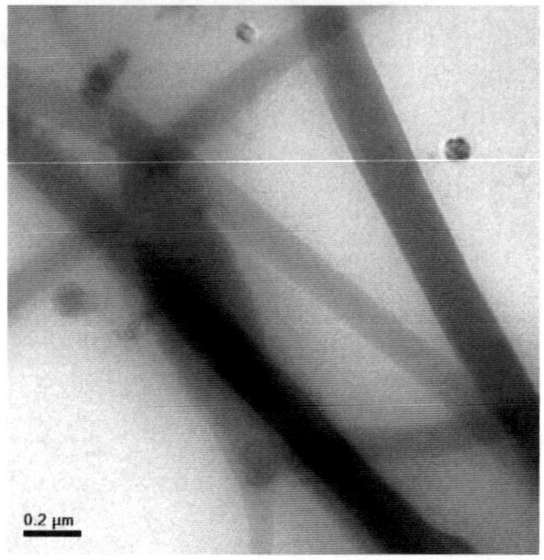

Figure 5.TEM of modified CNF composite

Figure 5 clearly shows that there is a better dispersion of CNF's in polymers when CNF's are functionalized. This is also attributes to the enhanced mechanical strength when plasma modified CNF were reinforced in the composites.

4. Conclusion

It has been noted that there is appreciable changes in the mechanical properties of composites after reinforcing modified CNF.TEM results also show that there is better dispersion of nano fibres in polymers after modification. FTIR results conclude significant enhancement in oxygen linkages after modification. These studies make it evident that modification of nano particles enhances the dispersion and interfacial adhesion.

Reference

[1] Andrews, Rodney, David J, Dali Q and Terry R 2002*Accounts of Chemical Research* **35-12** pp.1008-1017.
[2] Harris P J F 2004*International Materials Reviews* **49-1** pp.31-43.
[3] Lau K T, Shi S Q and Cheng H M 2003*Composites Science and Technology* **63-8** pp.1161-1164
[4] Thostenson E T, Ren Z and Chou T W 2001 *Composites science and technology*, **61-13** pp.1899-1912.
[5] Liu Y and Kumar S2014 *ACS applied materials & interfaces* **6-9** pp.6069-6087.
[6] Kumar S, Rath T, Mahaling R N, Reddy C S, Das C K, Pandey K N and Yadaw S B 2007 *Materials Science and Engineering: B*,**141-1** pp. 61-70.
[7] Bonin H W, Van Tine J R and Bui V T 2010.*Nuclear Technology*,**169-2** pp.150-179.
[8] Bonin HW, Hugues W and Miedema I 2008*Nuclear Technology* **164-2** pp.286-304.
[9] Weisenberger M C, Andrews R and Rantell T 2008*In Physical Properties of Polymers Handbook*pp. 585-598
[10] Leena G, Anirban S, Gomathi N, Joseph K, Neogi S and Reghunadan Nair CP 2015*Applied Surface Science*, **340** pp. 64-71.

IOP Conf. Series: Materials Science and Engineering **204** (2017) 012011 doi:10.1088/1757-899X/204/1/012011

Indigenous Indonesian Wild Silkworm Cocoon of Attacus atlas as Biocompatible Film Biomaterial

T G T Nindhia[1], Z Knejzlik[2], T Ruml[2], I W Surata[1] and T S Nindhia[3]

[1]Mechanical Engineering, Udayana University, Jimbaran, Bali, Indonesia, 80361
[2]Faculty of Food and Biochemical Technology, Technicka 5, 166 28 Prague Czech Republic
[3]Faculty of Veterinary MedicineUdayana University, Bali, Indonesia, 80000

E-mail: nindhia@yahoo.com, zdenek.knejzlik@vscht.cz, tomas.ruml@vscht.cz,

iwasura@gmail.com,snindhia@yahoo.com

Abstract. The biocompatible film made from wild silkworm cocoon of Attacus atlas is intorduced in this research for anticipation of demand on biocompatible film for regenerative medicine. The wild silkworm cocoon was indigenous Indonesia and was taken from it original location in Indonesia. Protocol for degumming method was obtained in this research by using treatment with NaOH solution at 0.1 M for 1 hour. The film was prepared by grinding the wet degummed fiber until pulp like state was obtained. The mixture was dropped in sequence on the hot ceramic plate with temperature around 60°C. The film thickness can be controlled precisely by using this technique. The film is soaked in alcohor for 1 day for stability testing and the result is found stable. The film is introduced in to COS-1 Cell suspension with previously washed in PBS solution and put in a chamber for biocompatibility testing. The cell are found able to grow and attach in first day observation and dramatically increase after 3 days observation. This is an indicate that the film that is produced from wild silkworm cocon of Attacus atlas has excellent biocompatibility.

1. Introduction

Diverse areas come together in an effort to maintain, improve, repair or replace biological function. Despite the remarkable of man-made synthetics biomaterial, their applications have been limited by challenges in biocompatibility, biodegradability and bioresorbability. Keys for improvement is lies in the development of functional materials that can interact with biological systems [1].

Virgin silks, is having an issues of biocompatibility if uses as sutures, with reactions from delayed hypersensitivity to acute and chronic process of inflammatory. Degummed silk incorporated directly into biomaterial architectures. The absence of the silk sericin that is released by degummed silk fiber, demonstrate minimum reaction of inflammatory, which enables successful implantation and cell culture. silk for biomaterial requires the separation of sericin from fibroin [1].

The biomaterial extracted from silk can be produce in the form of particle, fiber, hydrogels, scaffold, and film [1, 2]. Some silks already explored for biomaterial such as Bombyxmorifrom domesticated mulberry silkworm or from non-mulberry silk such as Antheraeamylitta, Antheraeaassamenis, Antheraeapernyi, Philosamiaricini, Samiacyn-thiaricini [1]. Other African source of non-mulberry wild silkworm cocoon also come in to attention like Gonometapostica, Epiphorabauhiniae, Anaphe

Content from this work may be used under the terms of the Creative Commons Attribution 3.0 licence. Any further distribution of this work must maintain attribution to the author(s) and the title of the work, journal citation and DOI.
Published under licence by IOP Publishing Ltd

panda, and Argemamimosae [3]. In Japan, The wild silk from species Samiacyn-thiaricini was well studied to obtain the fibroin from its cocoo[4].
In Korea, the biomaterial film was developed by using fibroin from Antheraeapernyi. The cocoon was degummed in solution of calcium nitrate, whased with mixture of surfactant that nonionic and sodum bicarbonate [5]
Wild silk also become attention for the researchers in the USA. Cocoon from wild insects such as H. euryalus, E. calleta, R. lebeau, A. oculae, H. gloveri and C. multifenestrata were investigated and degummed protocol was treating with ethylenediamine and sodium carbonate solution for 50 min at 80°C with the cocoon to solution ratio of 1:20 . The degummed silk was thoroughly washed and dried [6].
It is a need in recent year to provide film in medicine for drug release surface engineering. Layer by layer thin film assemblies was introduced because possible to control composition in precision and can be incorporate with snall molecules, heparin, peptides, protein and enzymes without activity lost. Nano scale thin coatings of silk fibroin could be formed by stepwise deposition using an aqueous process. The fibroin protein adsorbed onto substrates spontaneously and the thickness could be controlled by dipping solution and coating dehydration method. The silk protein locks in the coatings due to physical crosslinks of β-sheet crystals resulting in stable material coatings that do not cross linking reactions [7].
In this research, silk was obtained from the Cocoon of Atacus atlas which is Indigenous from Indonesia. Biofilm will be produced from this cocoon for biocompatible biomaterial.

2. Experimental

Indonesian indigenous wild silkworm cocoon of Attacus atlas was prepared for this purpose (Fig. 1). The solution of NaOH with concentration in the range of 0 - 1 M were prepared for the degumming process in 1 hour and the best result from that range was used for the degumming protocol of the cocoon. The result of degumming process was observed in detail by using scanning electron microscope (SEM). The fiber obtained from degumming process was cleaning by using water at 70°C.The hot ceramic plate with temperature around 60°C was prepared and the mixture was dropped in series on the hot plate to obtain film as desired thickness. For the purpose of sterilization, then the film was put in 70% ethanol for 24 hours. After that the film was washed with solution of PBS. The film was introduce in cultivation media of COS-1cell suspension for biocompatibility testing in the Chamber (humidity 95%, 5% CO_2, 37°C)for 1 day and 3 days testing. TheCOS-1 cellsarecellsofgreenmonkey kidney (African). This cell with vectorsthat requireSV40Tantigen, suitable fortransfection. The biocompatible characteristic (cell attachment and cell growth) were observed by using fluorescence microscope IX-71 (Olympus Japan). The sample should be transferred first in to PBS and washing it prior microscope observation.

(a) (b)

Figure. 1 Indigenous Indonesian wild silkmoth of Attacus atlas (a) and cocoon of wild silkworm Attacus atlas (b)

3. Result and Discussion

It is good to inform in this report that previously many degumming protocols are already establish for specific cocoon. The simplest method is by boiling [8]. Other method is by usingNa_2CO_3 [9]. It was reported that degumming can also be performed by usingurea that was developed because the level of degumming using boiling water is depending on the treatment time [10] .Degumming of the cocoon is performed traditionally by soap and can achieve optimum weight loss of at 30% [10].There was a study that the removal of sericin by using an enzymes (protease and lipase). Degumming by using enzyme involves the proteolytic degradation of sericin[10]. All the above mentioned protocol for degumming have been applied for cocoon of Attacus atlas and was found not success to release the fiber from the cocoon.

The degumming process with NaOH treatment then is introduced in this research to overcome the difficulty of degumming cocoon of Attacusatlas.Good result in releasing fiber from the bound in the cocoon is found with concentration 0.1M NaOH as can be seen at Fig 2. If more higher concentration to apply such as at 1 M NaOH, would made the cocoon completely soluble in to the solution. But if the low concentration to apply such as 0.01 M or 0.001 M the solution will not effect on the releasing of the fiber from the cocoon as well as jus by using water (0 M NaOH). The detail result of degumming with 0.1 M NaOH can be observed in Fig.3. Fine fiber is found released from the bound of the cocoon.

The film has been soaked in alcohol for 1 day and appears to be stable as can be seen in the Fig. 4. Same sample with different thickness that was obtained from multiple dripping at temperature about 60°C were also found very stable due to hydrophobic interactions. The films supported hMSC adhesion and proliferation [11]. The cells are found to attach andgrow onthe surface of thefilm that was made fromsilkworm cocoonsofAttacus atlas. The cell are well grow at the the 1st day (Fig. 5 a) and continuously grow as can be proofed at the 3rd day in the Fig. 5b .This is an indicationthat the biofilm made from wild silkworm cocoon of Attracus atlas has good compatibility.

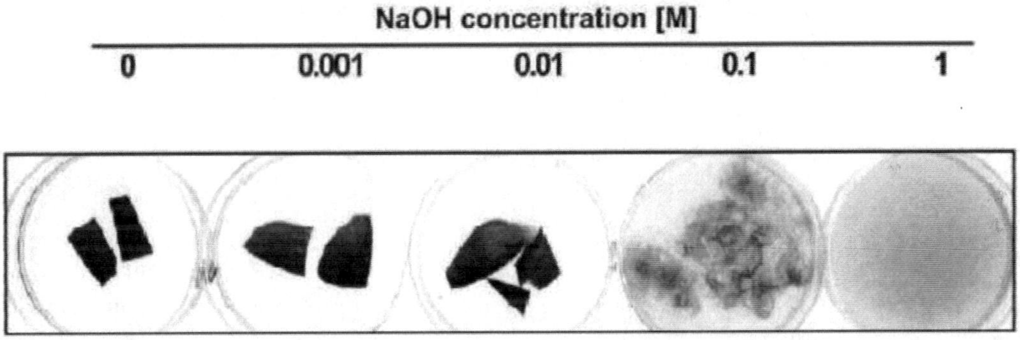

Figure 2 The Appearance of degumming process of cocoon of Attacus atlas at some variations concentration of NaOH.

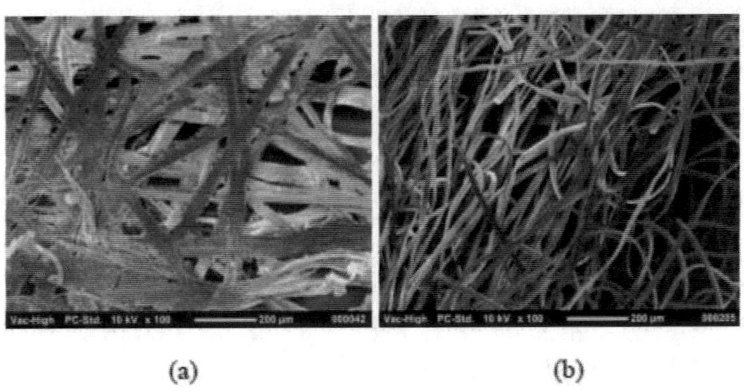

Figure 3 Texture fiber of cocoon Attacusatlass before degumming (a) and after degumming with 0.1M NaOH for 1 hour

Figure 4 The film has been soaked in alcohol for 1 day and appears to be stable

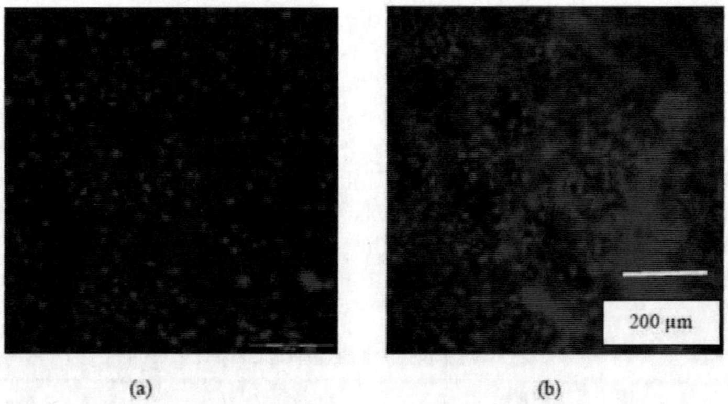

Figure 5 Cell are found well attached and grow on of the surface of the film in the 1st day (a). Remarkable cell growth in the 3rd day was observed on the surface of the film(b)

4. Conclusion

The Cocoon from wild silkworm cocoon of Attacus atlas can be processed for biocompatible film for medical application. The cocoon should be degummed by using 0.1M NaOH for 1 hour. The Film is produced by grinding the degummed fiber in wet condition and dripping in series on a hot ceramic plate at 60°C.The resulting film hasproved biocompatible and stable

Acknowledgment

This research was support financially from grants of thenon-tax revenues (PNBP) from the scheme of international collaboration research (*HibahPenelitianKerjasamaLuarNegri*) of Udayana University, Bali, Indonesia with collaborationwith theInstitute ChemicalTechnologyPrague, CzechRepublic

References

[1] Kundu B, Kurland NE, Bano S, Patra C, Engel FB, Yadavalli VK, Kundu SC: *Progress in Polymer Science*, 2014,**39**, pp. 251– 267

[2] KapoorS andKundu SC, *Acta Biomaterialia*,2016, **31**, pp. 17–32

[3] Teshome A, Vollrath F, Raina SK, Kabaru JM, and Onyari J, *International Journal of Biological Macromolecules*,2012, **50**, pp. 63– 68

[4] Asakura T, Yao J, Yang M, ZhuZ,and HiroseH,*Polymer*, 2007, **48** (2007), p. 2064,

[5] Kweon HY,Um IC, Park YH, *Polymer*, 2000 **41**, pp. 7361–7367

[6] Reddya N and Yanga Y, *International Journal of Biological Macromolecules*,2010, **46** pp.419–424

[7] Veparia C, and Kaplan DL, *Prog.Polym. Sci.*, 2007, **32**, pp. 991–1007

[8] Jiang P, Liu H, Wang C, Wu L, Huang J, and Guo C, *Material Letters*, 2005, **60**, pp. 919-925

[9] Meinel L, Fajardo R, Hofmann S, Langer R, Chen J, Snyde B, Vunjal-Novakovic G, and D. L. Kaplan DL, *Bone*, 2005, **37**, pp. 688-698

[10] Ho MP, WangH,Lau KT, Lee JH and Hui D, *Composites:Part B*, 2012, **43**, pp. 2801-2812

[11] Wang X, Hu X, Daley A, Rabotyagova O, Cebe P, and Kaplan DL 2007 *Journal of Controlled Release*, **121**,pp.190–199

IOP Conf. Series: Materials Science and Engineering **204** (2017) 012012 doi:10.1088/1757-899X/204/1/012012

Synthesis and Physico-Chemical Properties of Zinc Layered Hydroxide-4-Chloro-2-Methylphenoxy Acetic Acid (ZMCPA) Nanocomposite

N M Salleh[1], SM N Mohsin[2], S H Sarijo[1]and S A I S M Ghazali[3]

[1]Faculty of Applied Sciences, Universiti Teknologi MARA (UiTM), 40450, Shah Alam, Selangor, Malaysia
[2]Material Synthesis and Characterization Laboratory, Institute of Advanced Technology (ITMA), Universiti Putra Malaysia (UPM), 43400 UPM Serdang, Selangor, Malaysia
[3]Faculty of Applied Sciences, Universiti Teknologi MARA (UiTM), 72000, Kuala Pilah, Negeri Sembilan , Malaysia

E-mail: feezasalleh@gmail.com,sitihalimah@salam.uitm.edu.my,sheikhahmadizaddin@ns.uitm.edu.my

Abstract: Hybrid nanocomposite zinc layered hydroxide-4-chloro-2-methylphenoxyacetic acid (ZMCPA) was successfully synthesized via direct reaction of4-chloro-2-methylphenoxy acetic acid (MCPA) with zinc oxide (ZnO) under an aqueous environment. The pure phase and well-ordered ZMCPA was obtained when ZMCPA was synthesized at 0.3 M MCPA. The resulting nanocomposite, ZMCPA was characterized by powder X-ray diffraction (PXRD), fourier transform infrared(FTIR) spectroscopy, field emission scanning electron microscopy (FESEM), thermogravimetric and differential thermogravimetric analyses (TGA/DTG) and CHNS Analyzer. Basal spacing showed that the gallery height of zinc layered hydroxide (ZLH)expanded from 11.Å to 24.1 Å, indicating that the MCPA anions were successfully intercalated into the interlayer space of ZLH. The FTIR spectrum of the ZMCPA shows it resembles the spectra of ZnO and MCPA that confirms the intercalation of MCPA between the ZLH sheets. FESEM shows the ZnO precursor has granular structure and transformed into a flake –like structure when the nanohybrid is formed. . Percentage loading of MCPA interleaved in the nanohybrid is 50.2%.

Excellent intercalation properties of 2D layered material offer a new scope for developing hybrid materials at nanoscale dimensions or the so-called nanocomposite[1]. This brucite-like layered compound can be classified into two types from their structural anisotropy and chemical composition to produceintercalated compounds. Layered double hydroxide (LDH) and layered hydroxide salts (LHS). LDHcan be represented by a general formula, $[M^{2+}_{1\_x}M^{3+}_{x}(OH)_2] (A^{n-})_{x/n}YnH_2O$, where M^{2+} is a divalent cation (Ca^{2+}, Mg^{2+}, Zn^{2+}, Co^{2+},Ni^{2+}, Cu^{2+}, Mn^{2+}), M^{3+} is a trivalent cation (Al^{3+}, Cr^{3+}, Fe^{3+}, Co^{3+}, Ni^{3+},Mn^{3+}), A^{n}– is an interlayer anion such as (Cl^-, NO_3^-, ClO_4^-, CO_3^{2-}, SO_4^{2-}). The x value is the molar ratio, $M^{3+}/(M^{2+} +M^{3+})$. LHS can be represented with the general formula $M^{2+} (OH)_{2\_x}(A^{n}\text{–})_{x/n}.yH_2O$, where M^{2+} is the metallic cation (e.g.,Mg^{2+} , Ni^{2+} Zn^{2+} , Ca^{2+} , Cd^{2+} , Co^{2+} , and Cu^{2+}) and A is a counterion with ^{n}– charge[2]. LDH and LHS consists positively charged brucite-like inorganic layers and exchangeable anions and water molecules in the interlayer.Layered metal

Content from this work may be used under the terms of the Creative Commons Attribution 3.0 licence. Any further distribution of this work must maintain attribution to the author(s) and the title of the work, journal citation and DOI.
Published under licence by IOP Publishing Ltd

IOP Conf. Series: Materials Science and Engineering **204** (2017) 012012 doi:10.1088/1757-899X/204/1/012012

hydroxide (LMH) have various potential to be used in various industry such as ion exchanger [3], drug delivery system [4]and control release formulation [5].

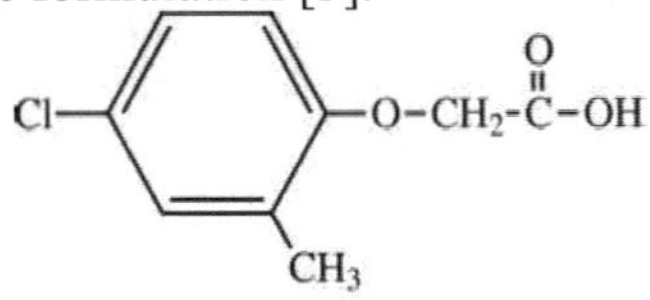

Figure. 1. Molecular structure of 4-chloro-2-ethylphenoxy acetic acid.

This paper reports our work on the intercalation of phenoxyherbicides namely 4-chloro-2-methylphenoxy acetic acid (MCPA) (Fig. 1) into ZnO via direct reaction of aqueous solution. MCPA is one of the derivatives phenoxyalkane carboxylic acid used as herbicide and a plant growth regulator. It is commonly used to control the broadleaf weeds, including thistle and dock, in cereal crops and pasture. It is selective for plants with broad leaves, and this includes most deciduous trees.

2. Experimental.

All solution was prepared using deionized water. MCPA (97%) was purchased from Aldrich, pure zinc oxide,ZnO(99%) and sodium hydroxide, NaOH (99 %) was purchased from R&M Chemical used without any further purification. MCPA was dissolved in 50ml of 90% ethanol into a beaker with a different concentration. Then, 0.2 g of ZnO was added into MCPA solution.The solution wastitrated with NaOH and stirred for 24h using amagnetic stirrer. The white precipitate was aged in oil bath shaker at 70 °C for 18h.After cooling, the resulting material was centrifuged, washed with deionized water and dried in an oven at 70 °C for 48h until fully dried. The resulting material was kept in the sample vial for further characterizations.

3. Characterization.

Power X-ray diffraction (PXRD) patterns were recorded with a XRD-6000(Shimazdu, Kyoto Japan) using Cu-Kα radiation (λ= 1.5418 Å) at 30kV and 30mA.The data was collected from 2 – 60° at a dwell time of 2°min^{-1}. Fourier Transform infrared (FTIR) spectra were recorded by a Perkin Elmer 1725x spectrophotometer in the range of 4000-450 cm^{-1}. The CHNS Analyzer 3400 Perkin Elmer Series II was used to determine the carbon and hydrogen percentage in the intercalation compound. The thermal decomposition behavior of the samples was carried out by using a thermogravimetric and differential thermogravimetric analyses (TGA/DTG) using a MettlerToledo instrument model TGA/SDTA851 thermogravimetric analyzer by heating rate at 10 °Cin the range of 35-1,000 °C under N_2 gas with a flow rate of 50 ml min^{-1}. Surface morphology of the samples was observed by a field emission scanning electron microscopy (FESEM) using a Zeiss Supra 40VP.

4. Results and Discussion

4.1Power X-ray diffraction.

Fig. 2(a) shows the PXRD patterns for ZnO, MCPA and ZMCPA nanohybrids synthesized at different concentration of MCPA. The PXRD patterns of ZnO indicate high crystallinity shows by a distinctive pattern of metal oxide with five sharp reflection peaks in the range of 2θ at 30-40°.However, comparing both0.2 and 0.3 M ZMCPA together, the nanolayered structure material obtained using 0.3 M ZMCPA, showsa symmetric peak with reflection up to 4 harmonics, respectively. This indicate the formation of the well-ordered 2D layered structure of the nanocomposite. Hence, the nanolayered structure material obtained from 0.3 M ZMCPA was selected for further characterizations and labeled as ZMCPA.The shifted diffraction peaks to the lower 2θ indicating an expansion of the inorganic layer host of zinc oxide layer.This confirmed the ZMCPA formation from simple reaction of ZnO with MCPA under aqueous environment was transformed into a layered hydroxides, and at the same time the intercalation of MCPA anionand the interlayer of the zinc-layered hydroxide took place.

IOP Conf. Series: Materials Science and Engineering **204** (2017) 012012 doi:10.1088/1757-899X/204/1/012012

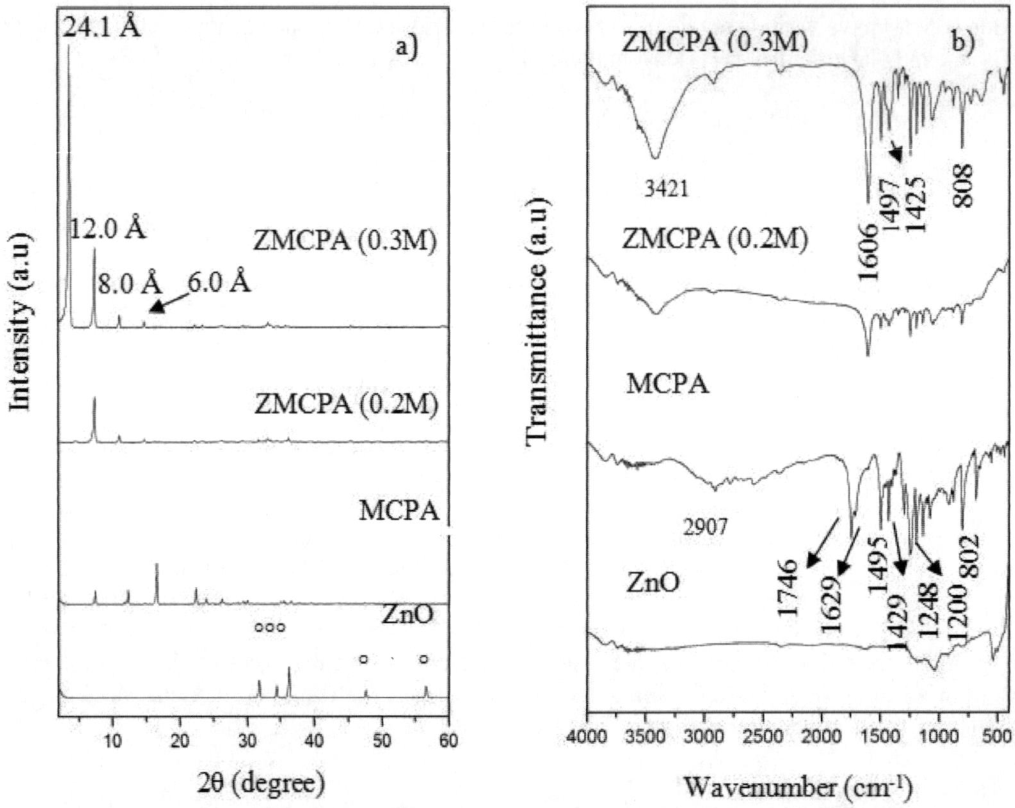

Figure. 2 PXRD patterns of ZnO, MCPA and ZMCPA nanohybrid (a) FTIR spectrum of ZnO, MCPA and ZMCPA nanohybrid (b) synthesized at different concentration of MCPA

4.2FTIR Spectroscopy.

Fig. 2(b) shows the FTIR absorption spectra of ZMCPA nanohybrids, MCPA anions and ZnO. For MCPA, a broad band at 2907 $cm^{-1}$is due to the O-H stretching vibration of the COOH. A strong and distinct band at 1746 cm^{-1} is ascribed to C=O stretching, 1629 cm^{-1}attributed to C=C vibration of the aromatic ring, 1495 $cm^{-1}$and 1429 $cm^{-1}$was due toO-H in-plane bending. The symmetric and asymmetric stretching mode of C-O-C vibration can be observed at 1247 and 1257 cm^{-1} and a band at 802 cm^{-1} is due to the C–Cl stretching vibration. ZnO displays a broad peak below 600 cm^{-1} which is the characteristic of Zn absorption bands [6]. The FTIR spectrum of ZMCPA nanohybrids shows a broad absorption band at 3421 cm^{-1} which corresponds to the O-H stretching vibration due to the absorbed and interlayer water molecules [7]. A strongband at 1606$cm^{-1}$is due to the C=C vibration of the aromatic ring.The band at 1497 and 1424 cm^{-1}·are due to the COO- stretching and C-O-C vibration of intercalated MCPA. At the lower wavenumber, the disappearance of the ZnO absorption bandfor the ZMCPA nanohybrid confirmed that all the ZnO was completely reacted with the MCPA in the aqueous solution, resulting in the formation of ZMCPA nanohybrid.This is evident by the FTIR spectra which shows combination spectra of both the inorganic host, ZnO and the guest anion, MCPA.This suggests that the MCPA anion was intercalated into the interlayer of the ZLH. Due to the intercalation process, some of the absorption bands shown were slightly shifted in position, due to the existence of interaction between the MCPA guest anion and the ZnO host as a result of the formation of new bonds.

4.3Elemental Analysis.

From the results, MCPA content 53.6 % of carbon (w/w) compared to 27.0 % of carbon (w/w) for the nanohybrid, ZMCPA. The high content of carbon in ZMCPA showed that MCPA anion was successfully intercalated between the interlayer of the layered material. The percentage loading of MCPA in ZMCPA nanohybrid estimated from the percentage of carbon is 50.2 %.

4.4Thermal Analysis.

The TGA/DTG thermogram of MCPA, ZnO, and ZMCPA are shown in Fig. 3. The TGA/DTG curve for MCPA shows a temperature maxima at 243.9°C with weight loss of 89.9%. This corresponds to complete decomposition of the organic compound MCPA. The TGA/DTG curve for ZMCPA shows a temperature maximaof309.5°C with weight loss of 52.6%. This indicates that MCPA encapsulated into the inorganic interlamellae is thermally more stable than their counterpart in the pure form. ZMCPA shows four intense of weight losses of 2.1 %, 52.6 %, 4.1 % and 33.9 % at temperature maxima of101.1°C, 309.5 °C,141.8°C, and 946.4°C, respectively. The first mass loss due to the removal of surface physisorbed and intercalated water molecules. The second mass loss is attributed to the decomposition of the intercalated MCPA. The third mass loss is attributed to the decomposition of the organic moiety in ZMCPA nanohybrid.

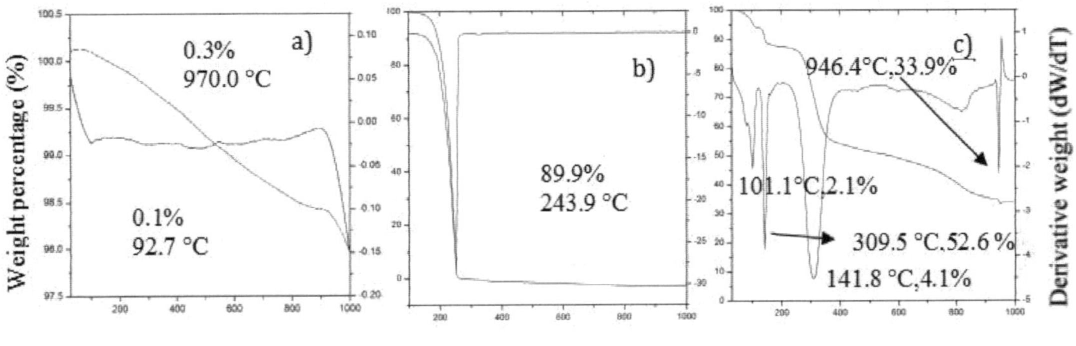

Figure. 3 TGA/DTG thermograms of the inorganic host ZnO (a), the guest anion, MCPA (b), and its nanohybrid, ZMCPA (c).

4.5Surface Morphology.

The surface morphologies of ZnO and ZMCPA are shown in Fig.4. ZnO shows granular structure with irregular particles of different sizes and various shapes at the nanometer scale. This structure was transformed into agglomerates of plate-like structure when the ZMCPA nanohybrid was formed by a direct reaction of ZnO with MCPA under an aqueous environment. This showed that the intercalation of MCPA into the interlayer of zinc-layered hydroxide resulted in a changed of surface morphology from well-defined hexagonal and square like structure to a plate-likestructure with a thickness in a nanometer range.

Fig.4 FESEM of ZnO (a), 0.2 M ZMCPA nanohybrid (b) and 0.3 M ZMCPA nanohybrid (c) at 100,000 x magnification.

5. Conclusion

An inorganic-organic nanohybrid of Zinc Layered Hydroxide 4-chloro-2-methylphenoxy acetic acid was successfully synthesized using a simple direct reaction of MCPA anion with ZnO under an aqueous environment with loading of MCPA interleaved in the nanohybrid of 50.2 %(w/w).

Acknowledgements
Funding for this research was provided by the Ministry of Higher Education (MOHE), under FRGS Grant 600-RMI/FRGS 5/3 (3/2014)and Graduan Insentif Postgraduate GIP 600-IRMI/GIP5/3 (0066/2016)Universiti Teknologi MARA (UiTM).

References

[1] M. Z. Hussein, N. S. S. A. Rahman, S. H. Sarijo, and Z. Zainal, "Herbicide-intercalated zinc layered hydroxide nanohybrid for a dual-guest controlled release formulation," *Int. J. Mol. Sci.*, **13**, no. 6, pp. 7328–7342, 2012.

[2] G. G. C. Arizaga, J. E. F. da C. Gardolinski, W. H. Schreiner, and F. Wypych, "Intercalation of an oxalatooxoniobate complex into layered double hydroxide and layered zinc hydroxide nitrate," *J. Colloid Interface Sci.*, **330**, no. 2, pp. 352–358, 2009.

[3] L. N. Sun and W. W. Tang, "Preparation of Layered Double Hydroxides Nanomaterails via the Anionic Surfactant," *Appl. Mech. Mater.*, **155–156**, pp. 907–911, 2012.

[4] B. Semete *et al.*, "In vivo evaluation of the biodistribution and safety of PLGA nanoparticles as drug delivery systems.," *Nanomedicine*, **6**, no. 5, pp. 662–71, Oct. 2010.

[5] L. P. Cardoso, R. Celis, J. Cornejo, and J. B. Valim, "Layered double hydroxides as supports for the slow release of acid herbicides," *J. Agric. Food Chem.*, **54**, no. 16, pp. 5968–5975, 2006.

[6] M. Z. Hussein, N. Hashim, A. H. Yahaya, and Z. Zainal, "Synthesis and characterization of [4-(2,4-dichlorophenoxybutyrate)-zinc layered hydroxide] nanohybrid," *Solid State Sci.*, **12**, no. 5, pp. 770–775, 2010.

[7] M. Z. Hussein, S. H. Al Ali, Z. Zainal, and M. N. Hakim, "Development of antiproliferative nanohybrid compound with controlled release property using ellagic acid as the active agent," *Int. J. Nanomedicine*, **6**, no. 1, pp. 1373–1383, 2011.

Synthesis of Superabsorbent Polymer via Inverse Suspension Method: Effect of Carbon Filler

MunirahEzzah Tuan Zakaria[1], SaidatulShima Jamari[1], Yeong Yi Ling[1], Suriati Ghazali[1,d]

[1]Faculty of Chemical and Natural Resources Engineering, Universiti Malaysia Pahang,26300, Gambang, Kuantan, Pahang, Malaysia

E-mail: munirahezzah@gmail.com, sshima@ump.edu.my, yyl1196@gmail.com, suriati@ump.edu.my

Abstract. This paper studies on the effect of the addition of carbon filler towards the performance of superabsorbent polymer composite (SAPc). In this work, the SAPc was synthesized using inverse suspension polymerization method. The process involved two different solutions; dispersed phase which contains partially neutralized acrylic acid, acrylamide, APS and NN-Methylenebisacrylamide, and continuous phase which contains cyclohexane, span-80 and carbon filler (at different weight percent). The optimum SAPs and filler ratio was measured in terms of water retention in soil and characterized by Mastersizer, FTIR and SEM. Biodegradability of the polymer was determined by soil burial test and SAPc with 0.02% carbon has highest biodegradability rate. SAPc with 0.04wt% carbon showed the optimal water retention percentage among all the samples. The synthesized SAPc producing spherical shapes with parallel alignment due to the addition of carbon fiber. It can be concluded that the addition of carbon fiber able to enhance the performance of the SAP composite (SAPc).

1. Introduction

Superabsorbent polymers (SAP) are a type of three-dimensional matrix constituted by branched or linear hydrophilic polymers that are chemically or physically crosslinked which possess the ability to absorb large quantities of water or biological fluids [1,2]. The crosslinked structure of this polymer allows it to maintain a stable network even in its swollen state[2]. Acrylic monomers or polymers, polyacrylamide and other polyacrylates are the most widely chosen as the raw materials to form high performance superabsorbent materials[3-9]. SAP can also be associated with another class of compound or substances resulting in a composite superabsorbent polymers. In this study, superabsorbent polymer has been incorporated with fibril substance which is carbon fiber. Interaction between polymer matrix and carbon fiber may lead to the enhancement of the polymer properties. Recently, demand for biodegradable polymer material increase dramatically as people worldwide

Content from this work may be used under the terms of the Creative Commons Attribution 3.0 licence. Any further distribution of this work must maintain attribution to the author(s) and the title of the work, journal citation and DOI.
Published under licence by IOP Publishing Ltd

starting to pays attention on saving the environment. Carbon fiber associated in superabsorbent polymer as filler not only can enhance the performance of the material when this polymer used as soil enhancer in agricultural field but also increase its biodegradability[6]. Oil palm or *Elaeisguineensis* is a commercial crop that are widely planted in Malaysia. Extracting crude oil from fresh oil palm fruit leaving abundant of fibrous empty fruit bunches (EFB). EFB is a natural fiber that typically consist of 24-65% cellulose, 21-34% hemicellulose and 14-31% lignin[11-16]. As it is very abundant and can widely be found in Malaysia, this fiber is the most suitable fiber used and converted into carbon fiber through hydrothermal carbonization process in this study. This study is carried out to investigate the effect of carbon filler towards superabsorbent polymer.

2. Experimental

2.1 Materials.

Oil Palm empty fruit bunch (EFB) was obtained from Lepar Oil Palm Mill, Malaysia. Chemicals used in this experiment were acrylic acid, N,N'-methylenebisacrylamide (NNMBA), ammonium persulfate(APS), sorbite anhydride monostearic acid ester (span-80), acrylamide, and cyclohexane which were supplied by Merck. Carbon fiber is obtained by processing OPEFB fiber through hydrothermal carbonization process.

2.2 Methods

2.2.1 Synthesis of Superabsorbent Polymer Composites (SAPc) Via Inverse Suspension Polymerization.

Superabsorbent polymer composites was prepared by mixing cyclohexane and span-80 in a flask then heated to 55°C with nitrogen purged for 15 minutes. Carbon fiber (0-0.04wt%) was then added into the solution and left stirred for 10 minutes. Partially neutralized acrylic acid, acrylamide, initiator APS and cross linker NNMBA was mixed in another beaker. The mixture was then added slowly into the flask and the reaction held for 3h at 65 °C with 300 rpm agitation speed. The resulted polymer was filtered, washed and dried in the oven.

2.2.2 Characterization.

The superabsorbent polymer composite (SAPc) were characterized by recording the spectrum on a Fourier Transformed Infrared Spectroscopy (FTIR) by using the attenuated total reflectance technique(ATR). The morphologies of this polymer were analyzed using Scanning Electron Microscope

2.2.3 Measurement of Particle Size.

The measurement of the particle size of the SAP and SAPc beads was determined using a particle analyzer, MastersizerSrirocco 3000 model.

2.2.4 Water Retention in Soil.

The soil was sieved by using sieve shake(2mm)and dried in the oven at 60°C. Then, 50g of soil was inserted into the container and mix with 0.5wt%, 1.5wt% and 2.5wt% of SAP and SAPc. The water was added until all the soil in the container is wet. The container was then weighed using weighing balance to determine its initial weight, Mo. The final weight, Mi was recorded each day for the following 30 days and water retention is calculated as ratio of final and initial weight; Mi / Mo.

2.2.5 Soil Burial Test (Biodegradability).

The soil was sieved using sieve shaker (2mm) and is dried in the oven at 60°C. 0.08g of SAP and SAPc inserted into a teabag and placed in containers in which the soil covered the SAP and SAPcwholely. The container was covered with aluminium foil and kept at room temperature. All samples were weighted with consistent time interval (5 days) and weight loss percentage was calculated as (initial weight at the beginning – final weight after ten days)/ initial weight at the beginning x 100%.

3. Result and Discussion

3.1 Water Retention in Soil.

Figure 1 and 2shows water retention percentage of SAP and all SAPc samples varies in different sets of weight percentage (0.5wt%, 1.5wt% and 2.5wt%). SAPc samplethat contains 0.04wt% of carbon shows the highest water retention percentage from all sets of weight percentage which were 70.06% for 0.5wt% set, 70.07% for 1.5wt% set and 70.35% for 2.5wt% set.

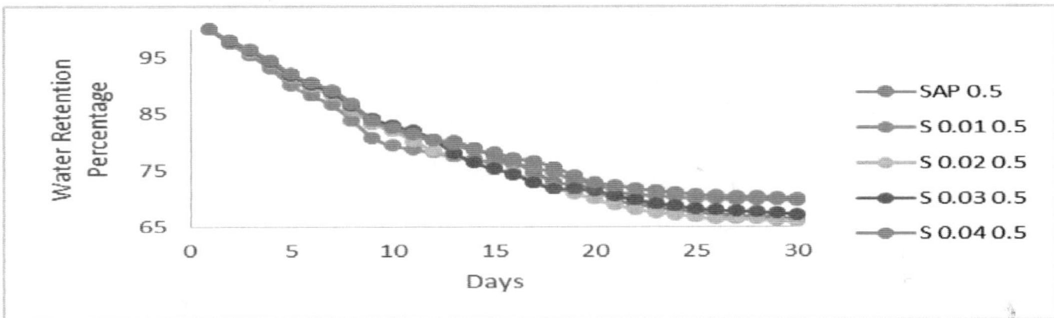

Figure 1.Water Retention of SAP and SAPc samples with 0.5wt% in soil.

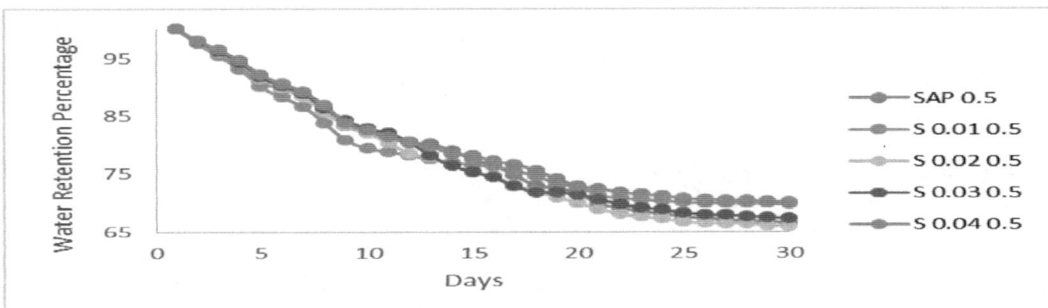

Figure 2.Water Retention of SAP and SAPc samples with 1.5wt% in soil.

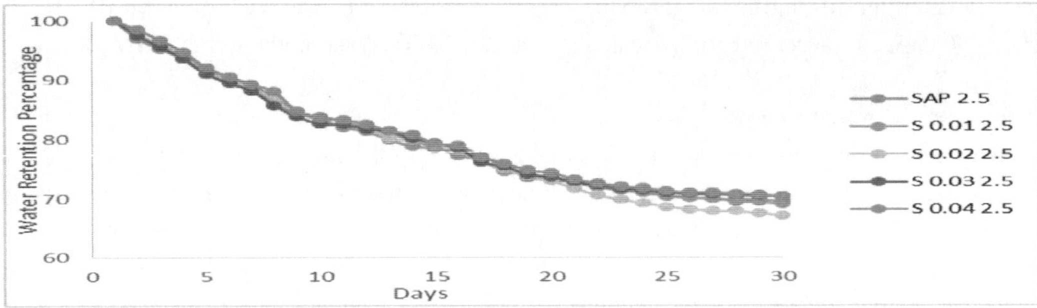

Figure 3.Water Retention of SAP and SAPc samples with 2.5wt% in soil.

3.2 CharacterizatiOon.

Figure 4 represent the FTIR spectrum of SAP and SAPc samples. A strong and broad OH−stretch with H−bonded at the band of 3250 cm^{-1} indicates that there is presence of alcohols or phenols group. Medium peak of 2835 cm^{-1} and 1450cm^{-1} indicates C−H with alkanes group are formed whereas at the peak of 1650 cm^{-1}, there is medium N−H bond which signifies the presence of primary amines. Then, medium C−N stretch from aliphatic aminesdetected at 1055cm^{-1} peak. A broad and strong peak of −C≡−C−H detected at 650 cm^{-1}.

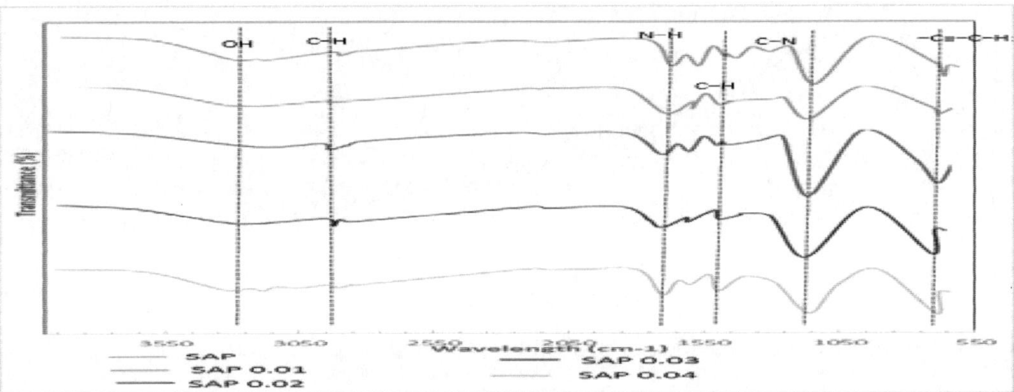

Figure 4.FTIR spectrum of SAPc

Figure 5 shows Scanning Electron Microscope micrograph of SAPc that indicates that the SAPc produced via inverse suspension polymerization was in beads and spherical form.

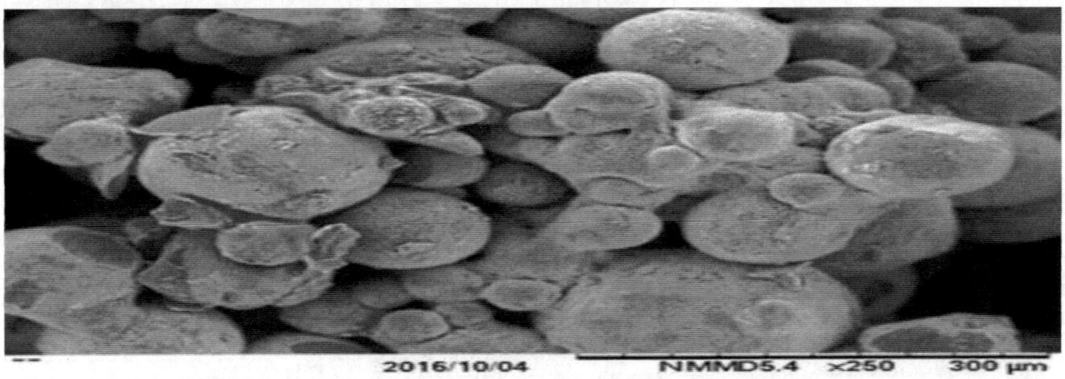

Figure 5. SEM micrograph of SAPc at 250 times of magnification.

IOP Conf. Series: Materials Science and Engineering **204** (2017) 012013 doi:10.1088/1757-899X/204/1/012013

3.3 Soil Burial Test (Biodegradability).

Soil burial test was the test used to determine the biodegradability of SAP and SAPc. In this test, the biodegradability was determined by the percentage of polymer weight loss over time. Figure 6 shows SAPc with 0.02wt% carbon possessed the highest percentage of weight loss over time. SAP sample records lower percentage of weight loss than SAPc with 0.02wt% but higher than other SAPc samples. Presence of carbon as filler does prove that carbon filler improved the biodegradability of SAP as shown by SAPc sample with 0.02wt% . However, the other SAPc samples shows lower degradability rate may be because too much carbon filler affect the polymerization of SAPc.

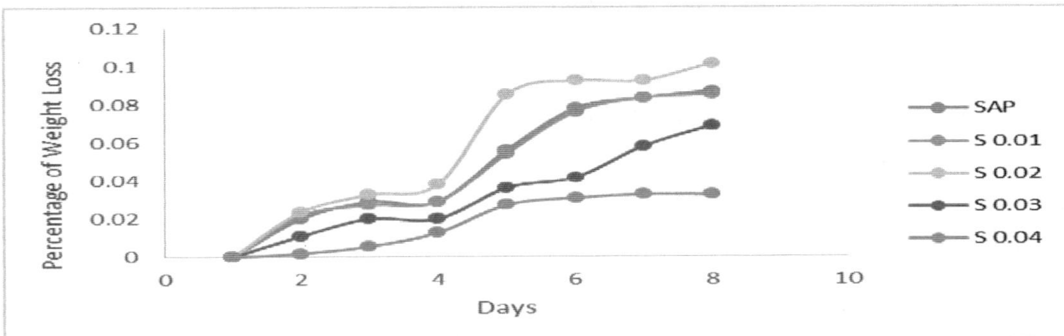

Figure 6. Percentage of weight loss of SAP and SAPc

3.4 Particle Size Analysis.

The size of SAPare relatively smaller thanSAPc. SAP particle size range from 563.667µm to 709.627 µm. Meanwhile, SAPcparticle size range from 893.367 µm to 1124.683 µm. Thus,carbon filler added into the SAP polymer significantly affect the size of superabsorbent polymer.

4. Conclusion.

SAPc with 0.04wt% portrayed the highest water retention rate concludes that more carbon percentage helps SAPc to retain more water than original SAP. However, SAPc with 0.02% carbon possessed the highest degradability rate than other SAP and SAPc proves that carbon filler improved SAP biodegradability. Overall, SAPc possessed better performance than normal SAP. This polymer will be further studied and applied in agricultural field.

5. Acknowledgement.

The authors would like to be obliged to Universiti Malaysia Pahang for providing laboratory facilities and financial assistance under project no. RDU140106. Special thanks also to LCSB Sdn. Bhd. for the raw material provided.

6. References

[1] Chang C, Duan B, Cai J, Zhang L, Superabsorbent hydrogels based on cellulose for smart swelling and controllable delivery, *Eur. Polym. J.***46** (1)(2010) pp 92–100.

[2] Pourjavadi A, Harzandi A M,Hosseinzadeh H, Modified carrageenan 3. Synthesis of a novel polysaccharide-based superabsorbent hydrogel via graft copolymerization of acrylic acid onto kappa-carrageenan in air, *Eur. Polym. J.***40** (7) (2004) pp 1363–1370

[3] Laftah W A, Hashim S, Ibrahim A N, Polymer hydrogels: a review, *Polym. Plast. Technol. Eng.***50** (14) (2011) pp 1475–1486.

[4] Rodrigues F H A, Spagnol C, Pereira A G B, Martins A F, Fajardo A R, Rubira A F, Superabsorbent hydrogel composites with a focus on hydrogels containing nanofibers or nanowhiskers of cellulose and chitin, *J. Appl. Polym. Sci.***131** (2) (2014).

[5] Omidian H, Rocca J G, Park K, Advances in superporous hydrogels, *J. Control. Release***102** (1) (2005) pp 3–12.

[6] Kazanskii K S, Dubrovskii S A, Chemistry and physics of agricultural hydrogels, *Adv. Polym. Sci.***104** (1992) pp 97–133.

[7] Mekonnen T, Mussone P, Khalil H, Bressler D, Progress in bio-based plastics andplasticizing modifications, *J. Mater. Chem. A 1***43**(2013) pp 13379–13398.

[8] Rinaudo M, Chitin and chitosan: properties and applications, *Progr. Polym. Sci.* **31**(7)(2006) pp 603–632.

[9] Heinze T, Liebert T, Heublein B, Hornig S, Functional polymers based on dextran,*Adv. Polym. Sci.* **205** (2006) pp 199–291.

[10] Zhao Y, Kang J, Tan T, Salt-, pH- and temperature-responsive semi- interpenetrating polymer network hydrogel based on poly(aspartic acid) and poly(acrylic acid),*Polymer***47** (22) (2006) pp 7702–7710.

[11] Jamari S S, Howse J R, The effect of the hydrothermal carbonization process on palm oil empty fruit bunch, *Biomass and bioenergy* **47**(2012) pp 82 – 90.

[12] Sumathi S, Chai S P, &Mohamed A R,Utilization of oil palm as a source of renewable energy in Malaysia. *Renewable & Sustainable Energy Reviews*, **12**(2008), pp 2404–2421.

[13] Dorez G, Ferry L, Sonnier R, Taguet A, Lopez-Cuesta J M, Effect of cellu- lose, hemicellulose and lignin contents on pyrolysis and combustion of naturalfibers, *J. Anal. Appl. Pyrol.* **107** (2014) pp 323–331.

[14] Lee T, Ooi C H, Othman R, Yeoh F Y, Activated carbon fiber—the hybrid of carbon fiber and activated carbon, *Rev. Adv. Mater. Sci.***36** (2014) pp 118-136

[15] MPOC. A summary on the performance of the Malaysian oil palm industry e 2009; 2010.

[16] Wahid M B. Foreword:overview of the Malaysian oil palm industry 2008. Available from: http://econ.mpob.gov.my/ economy/EID_web.htm; 2009.

[17] Alam M Z, MuyibiA S, Mansor M F, Wahid R, Activated carbons derived from oil palm empty-fruit bunches: application to environmental problems. *J Environ Sci* **8**(2007) 19:103.

[18] Zulkiple N, Maskat M Y, & Hassan O, Pretreatment of oil palm emptyfruit fiber (OPEFB) with aqueous ammonia for high production of sugars. *Procedia Chemistry*, **18**(2016), pp 155–161.

[19] SuriyaPalamaea, PongsathornDechatiwongseb, WannaChoorita, Yusuf Chistid, PoonsukPrasertsane, Cellulose and hemicellulose recovery from oil palm empty fruit bunch(EFB) fibers and production of sugars from the fibers,*Carbohydrate Polymers***155** (2017) pp 491–497.

[20] FideliaNnadi and Chris Brave. Environmentally friendly superabsorbent polymers for water conservation in agricultural lands. *Journal of Soil Science and Environmental Management Vol. 2,***7**(2011): pp 206-211

[21] Wu L, Liu M and Liang R, Preparation and properties of a double-coated slow-release NPK compound fertilizer with superabsorbent and water-retention,*Bioresource Technology* **99** (2008): pp 547–554

IOP Conf. Series: Materials Science and Engineering **204** (2017) 012014 doi:10.1088/1757-899X/204/1/012014

Fabrication of a silica aerogel and examination of its hydrophobic properties via contact angle and 3M water repellency tests

Z. Mazrouei-Sebdani[1], L. Javazmi[2], A. Khoddami[1], F. Shams-Ghahfarokhi[1] and T. Low[2].

[1]Department of Textile Engineering, Isfahan University of Technology, Isfahan 84156-83111, Iran
[2]Department of Mechanical and Electrical Engineering, University of Southern Queensland, Toowoomba QLD 4350, Australia

E-mail: z.mazrouei@tx.iut.ac.ir, leila.javazmi@usq.edu.au, khoddami@cc.iut.ac.ir, farzane2906@gmail.com, tobias.low@usq.edu.au

Abstract. Aerogels are dry gels with a very high specific pore volume. Aerogels with increased hydrophobicity have significant potential to expand their use as lightweight materials. Considering its special nanostructure and exceptional properties, this paper focuses on the synthesis and hydrophobic evaluation of a silica aerogel. The structural properties were investigated by measuring density, SEM micrographs, and BET analyses. Also, the hydrophobic evaluation was carried out by measuring 3M water repellency and water/alcohol contact angle. The BET analysis showed successful synthesis of the nanoporous silica aerogel with a pore size of 24 nm and porosity of 89%. The synthesized aerogel showed 3M water repellency of 3 and water contact angle of 129.6°. Also, it is worth-mentioning that as the alcohol content of the drops in 3M water repellency test is increased, the drop contact angle is decreased due to its lower surface tension. Thus, the contact angle reaches the zero at 3M water repellency test number of 4 (water/alcohol 60/40).

Introduction

The word aerogel was firstly presented by Kistler in 1932 to title gels in which the liquid was replaced with a gas, without collapsing the gel solid network [1]. Aerogels are dry gels with a very high specific pore volume. Aerogels, known as frozen smoke or air-glass, are contained of nanoparticles with usual dimensions of less than 10 nm and pore sizes of less than 50 nm in diameters [2-4]. If aerogels are sensitive to moisture, it causes them to lose their structural network integrity during their application. Thus, aerogels with increased hydrophobicity have significant potential to expand their use as lightweight structural, insulating or shock absorbing materials especially in aeronautics, microelectronics, and sensing applications.

There are three primary techniques used to make hydrophobic silica aerogels which involve vapor phase after treatment, co-precursor techniques, and derivatization methods [5].

The majority of the work reported in the literature has focused on the use of organosilanes specially methytrimethoxysilane (MTMS) or trimethylethoxysilane (TMES) as a co-precursor to make hydrophobic gels. They achieved a water contact angle up to 180°. For example, Standeker et al. [6] studied the effects of MTMS and TMES on hydrophobicity and achieved contact angles of 42–173° for MTMS/TMOS molar ratios of 0.5–5 and contact angles of 100–180° for TMES/TMOS molar ratios of

Content from this work may be used under the terms of the Creative Commons Attribution 3.0 licence. Any further distribution of this work must maintain attribution to the author(s) and the title of the work, journal citation and DOI.
Published under licence by IOP Publishing Ltd

0.5–5. Additional studies were performed by Rao and coworkers [7] in which the monolithic aerogels with contact angle up to 140° were fabricated. On the other hand, silylation especially with (trimethylchlorosilane) TMCS has been performed using a variety of precursor materials to form hydrophobic gels which are able to use ambient drying. For example, Jeong et al. [8] provide a direct comparison of TEOS-based, ambient-dried aerogel powders with and without silylation and show an increase in contact angle from near zero to 158 when silylation was used. Different from these works, no research has been done on the surface energy investigation of the aerogel material via 3M water repellency test.

This paper focuses on one of the most studied aerogel materials, silica aerogels and investigates its synthesis and hydrophobic properties containing surface energy in a new way.

Experimental

Aerogel synthesis. The schematic procedure for synthesizing the silica aerogel is shown in Fig. 1. According to fig.1, the silica sol was prepared with a waterglass or sodium silicate solution with a specific gravity of 1.39 which was diluted to 1:4 (v/v) with water. The sodium ions were exchanged with proton ones by going through an Amberlite IR 120 H ion exchange resin-filled column. The gelation was occurred via addition of the ammonia solution (1%) to the collected silicic acid with pH values of 2.4–2.7 in a plastic vessel. The wet gels were aged for 3 h to strengthen their structures. The aged hydrogel was immersed in the propone-2-ol, n-Hexane, and TMCS/n-Hexane (1:5 V/V) mixture for 24 h to be changed to alcogel, organogel, and modified organogel, respectively. The resulted modified gels were dried at room temperature for 24 h. The dried gel was heat treated at 50 and 230°C for 1 h to make the aerogel granules.

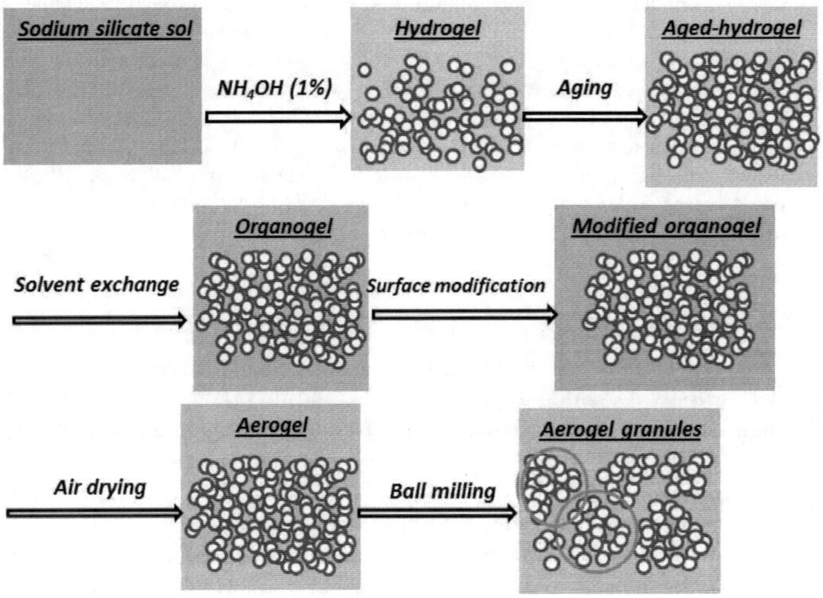

Figure. 1 A schematic of aerogel synthesis.

Hydrophobic properties. To study the level of the samples' hydrophobicity, both sliding angle and 3M water repellency [9] tests were examined. The samples were tested for water repellency using the water/alcohol drop test. The samples are placed flat on a smooth, horizontal surface. Beginning with the lowest numbered test liquid, three small droplets (approximately 5mm in diameter) are placed onto the sample using a pipette. The droplets are observed for 10 s. If after 10 s, two of the three droplets are still visible as spherical to hemispherical, the fabric passes the test. Samples are rated as pass or fail of the appropriate test liquid, W–10. The rating given to a sample is for the highest test liquid remaining visible after 15 s. 3M water repellency liquids number and content is indicated in Table 1. In general, water repellency rating of 2 or greater is desirable. The location of a droplet on the surface of the hydrophobic treated samples allows evaluating the water contact angle. The images were taken using a USB digital microscope, Digi Micro Scope. The angles between the liquid/solid and liquid/vapor interfaces were measured using the Digimizer software.

Table 1 3M water repellency liquids number and content.

Test number	Water/ alcohol ratio
0 (water)	100/0
1	90/10
2	80/20
3	70/30
4	60/40
5	50/50
6	40/60
7	30/70
8	20/80
9	10/90
10 (alcohol)	0/100

SEM and BET. The SEM images were obtained using a KYKY-EM3200 microscope.
The BET analysisto evaluate the specific surface areas, total pore volume, and mean pore diameter and distribution was carried out using a Belsorp mini IIapparatus (ISO 9277).
Density and porosity. The porosity of the manufactured silica aerogel was calculated using Eq. 1[9].

Where, P, ρ_b, and ρ_s, represent the porosity (%), bulk density (g/cm^3), and skeletal density of the amorphous silica (g/cm^3), respectively. The bulk density and total pore volume were calculated by Eq. 2.

Where, m is the mass (g), V_α is the apparent volume (cm^3), and v is the total pore volume (m^3/g).

Result and discussion

Aerogel synthesis. Fig. 2 shows the primary hydrogel before and the resulted aerogel granules. As shown in Fig. 2, the initial transparent gels were changed to the opaque hydrophobic aerogel during the surface modification process. Two phases containing of water/HCl phase under the *n*-Hexane phase, obtained after successful treatment of the gel with TMCS is indicated in Fig. 2 (B). The hydrophobic modified gel was floated in the *n*-Hexane phase. This phenomenon is explained by the following main reaction (1) that can occur during the solvent exchange/modification of the wet gel [10].

$$\text{TMCS} + {\equiv}\text{Si}-\text{OH} \rightarrow {\equiv}\text{Si}-\text{O}-\text{Si(CH}_3)_3 + \text{HCl} \qquad (1)$$

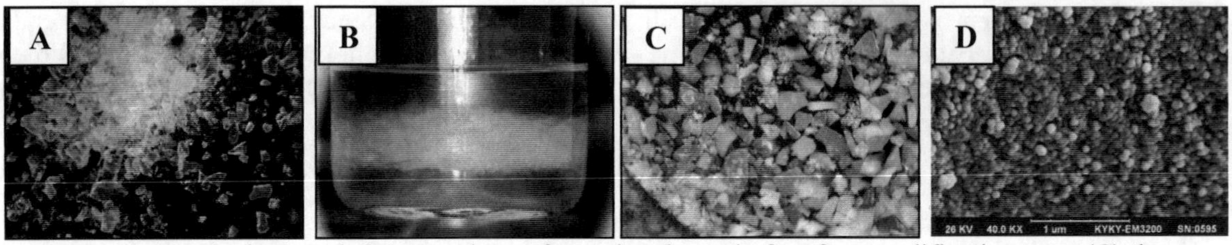

Figure. 2 (A) The hydrogel, (B) two phases formed at the end of surface modification step, (C) the aerogel granules, and (D) SEM micrographs

Also, the structure of the synthesized aerogel in a nanoscale form is shown in a SEM micrograph in Fig. 2 (D).

Table 2 shows the structural parameters of the synthesized aerogel obtained by the BET analysis. It is clear that the porous powder is successfully synthesized with a surface area of 815 m^2/g, a mean pore diameter of 24 nm, and a total pore volume of 3.5 m^3/g. Thus, the obtained dried gel in this work can be called an aerogel because all aerogels have nano-pores with diameters of less than 50 nm [11].

Table 2 Aerogel characteristics

Density (g/cm^3)	Total pore volume (m^3/g)	Mean pore diameter (nm)	Porosity (%)	Water contact angle (°)	3M water repellency
0.25	3.5	24.0	89	129.6	3

Hydrophobic properties. Surfaces are known to interact with their environment in a variety of ways. One central concept governing such interactions is the minimization of the interfacial surface-free energy [12]. The resulted aerogel shows the 3M water repellency value of 3. It indicates the silylation of the OH groups to the O–Si $(CH_3)_3$ groups with TMCS. So, it can be concluded that TMCS successfully reacted with Si–OH groups which are the main source of hydrophilicity of the aerogels. Water/alcohol drops, used in 3M water repellenct test, on the aerogel surface is shown in fig. 3. Also, water/alcohol drops contact angles on the aerogel surface is indicated in Table 3. The aerogel granules exhibit a water contact angle usually associated with that of high hydrophobic gels modified by the TMCS hydrophobic agent.

Also, there is an indirect relation between the water/alcohol number and its contact angle as shown in fig. 3 and Table 3. Thus, as alcohol content of the test droplet is increased, its contact angle is decreased until it reached the Zero at test number of 4 (water/alcohol 60/40). This is due to this fact that the addition of alcohol to the water phase decrease it surface tension leading to easier penetration of the drops to the surface.

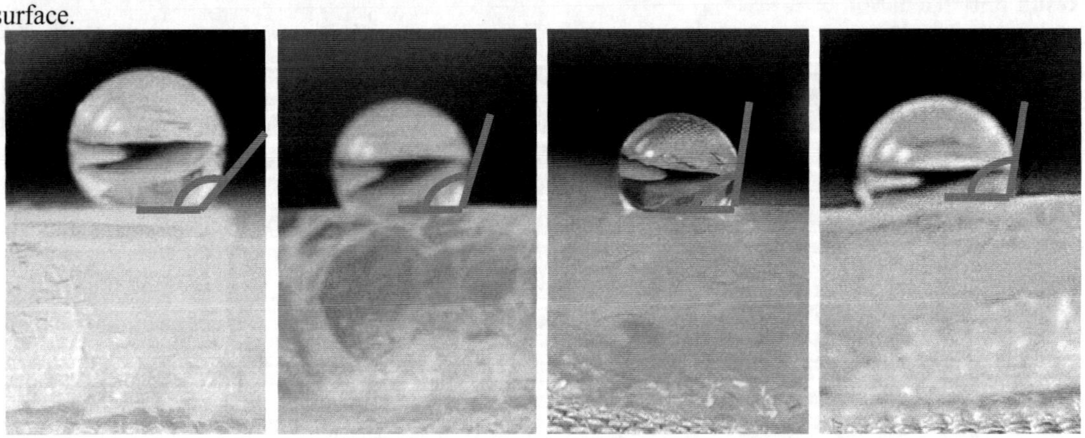

Figure. 3 3M water repellency solutions. (A) 100/0, (B) 90/10, (C) 80/20, (D) 70/30 water/alcohol drops on the aerogel surface

Table 3 Contact angle of water/alcohol drops used in 3M water repellency test on the aerogel surface

Test number	W	1	2	3	4
Contact angle	129.6	126.3	118.1	108.2	0

For better comparison, the colour water drop and 60/40 water/alcohol drop on the aerogel surface is indicated in fig. 4. It is clearly obvious that the 60/40 water/alcohol drop spreads on the aerogel surface and wet it completely.

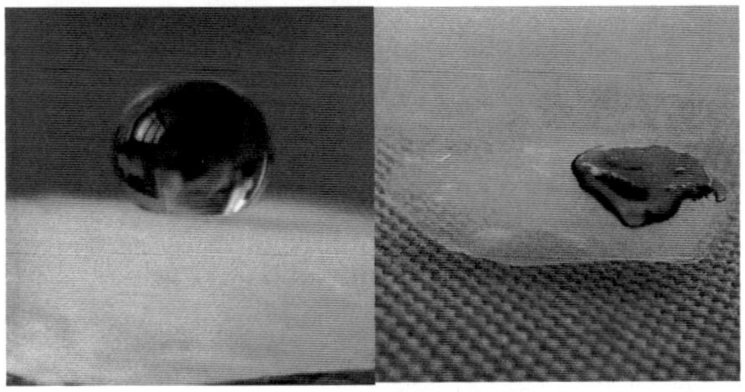

Figure. 4 (A) A (100%) water droplet, and (B) (60/40) water/Isopropan 2-ol drop on the synthesized aerogel surface.

Conclusion

In this research, waterglass based aerogel granules were prepared by a sol–gel process and ambient pressure drying. According to the results, the silica aerogel was fabricated with a mean pore diameter of 24 nm, atotal pore volume of 3.5 m^3/g, and porosity of 89%. The silica aerogel showed hydrophobic properties by a water contact angle of 129.6° and 3M water repellency of 3 with water/alcohol mixture with volumetric ratio of 70/30. Thus, the liquids with surface energy of less than 3M water repellency value of 4 cannot be absorbed on the aerogel surface which can be investigated in further studies.

Acknowledgment

Financial support of the Isfahan University of Technology (IUT) is gratefully appreciated.

References

[1] S. Kistler: *J. Phy. Chem.* **36** (1932), p. 52.
[2] S. Yun, H. Luo and Y. Gao:*RSC Adv.* **4** (2014), p. 4535.
[3] S.H. Hyun, J.J. Kim and H.H. Park: *J. Am. Ceram. Soc.* **83** (2000), p. 533.
[4] L. Zhong, X. Chen, H. Song, K. Guo and Z. Hu: *New J. Chem.* **39** (2015) p. 7832.
[5] M.A. Aegerter, N. Leventis, M.M. Koebel: *Aerogels handbook* (Springer, New York 2012).
[6] S. Standeker, Z. Novak and Z. Knez: *J. Colloid Interface Sci.* **310** (2007), p. 362.
[7] A.V. Rao and D. Haranath: *Microporous Mesoporous Mater.* **30** (1999), p. 267.
[8] A.Y. Jeong, S.M. Koo and D.P. Kim: *J. Sol-Gel Sci. Technol.* **19** (2000), p. 483.
[9] Z. Mazrouei-Sebdani, A. Khoddami and S. Mallakpour: *Colloid Polym. Sci.* **289** (2011), p. 1035.
[10] C. Lee, G. Kim and S. Hyun: *J. Mater. Sci.* **37** (2002), p. 2237.
[11] M. Linden, S. Schacht, F. Schuth, A. Steel and K.K. Unger: *J. Porous Mater.* **5** (1998), p. 177.
[12] A. Khoddami, Z. Mazrouei-Sebdani, and S. Mallakpour: *J. Text. Polym.***1** (2013) p. 36.

Polypropylene/olive pit & almond shell polymer composites: wear and friction

M Tasdemir

Marmara University, Tech. Faculty, Metallurgy and Mat. Eng. Dep. Istanbul, 34722, Turkey

E-mail: munir@marmara.edu.tr

Abstract: Wood plastic composites (WPC) are made from wood and annual plant fiber or flours, mixing with plastics materials. WPC provide better properties than resources that form it. This renewable material has many utilization areas because of outstanding properties such as enhanced strength, stiffness, creep, physical and mechanical properties and dimension stability. In the present work, series of filled Polypropylene (PP) composites with olive pit and almond shell flour loading (between 0–40 wt %) were prepared, to study the effect of the filler content on the mechanical, wear and friction properties of polypropylene polymer composites.

1. Introduction

Particularly, researches with regard to wood plastic composite materials have been continuously increased within the last 20 years because of the advantageous characteristics of each constituent in composite materials [1]. It is believed that source of petroleum based products are limited and uncertain. So an alternative with cheap sustainable and easily available raw material is required. The countries growing plant and fruit are not for only agricultural purpose but also to generate raw materials for industries. Most of the developing countries trade lignocellulosic fibers for improving economic condition of poor farmers as much as country support. Recently polymer composites containing cellulosic fibers are under focus in literature as well as industries [2]. The olive oil extraction processes generates huge amounts of solid lignocellulosic waste, representing 30-35% by weight of the olive fruit. The chemical composition of which; mainly composed of lignin, hemicellulose and cellulose. Olive waste has a growing potential as a reinforcement filler materials among many other uses [3-8]. Qutaiba studied of a new range of sustainable reinforced polymer composite materials using powdered olive pits as a novel filler material to be used with epoxy resin. Also the influence of the untreated and treated powder loading (weight fraction) on the void content and the mechanical properties of the composites was examined. The results show significant improvements in mechanical properties for composites reinforced with treated olive pits than composites reinforced with untreated olive pits [9]. An investigation was carried out on the effects of olive pit and almond shell powder ratio on the mechanical, wear and friction properties of polypropylene polymer composites. Olive pit and almond shell powder, in four different concentrations (10, 20, 30 and 40 wt %), was added to PP to produce composites. The properties, including density, water absorption, Vicat softening point, heat deflection temperature, wear rate, friction of the composites were investigated.

2. Experimental

Nine different polymer composites were prepared. Compositions of PP/olive pit and almond shell flour polymer composites that were formed are given in Table 1. PP (Moplen EP 3307) supplied by

Content from this work may be used under the terms of the Creative Commons Attribution 3.0 licence. Any further distribution of this work must maintain attribution to the author(s) and the title of the work, journal citation and DOI.
Published under licence by IOP Publishing Ltd

LyondellBasell. Its density is 0.900 g/cm^3. Olive pit and almond shell were pulverized (about 40 nm size) with Siemens simatic C7-621 control system device. Polypropylene (PP), olive pit and almond shell flour were dried overnight at 105 °C in a vacuum oven prior to melt blending. Mechanical premixing of solid compositions was done using a LB-5601 liquid-solids blender for 15 min. Samples with various proportions of polymer composites were produced between 180-220 °C at 12 bar pressure, and a rotation rate of 28-32 rpm, with a Microsan extruder. Polymer composites were also dried in vacuum oven at 105 °C for 24 hours after extrusion. Subsequently, test samples were manufactured by injection molding. Injection temperature was 180-220 °C, pressure was 100-120 bar and screw speed was 20 rpm.

Table 1. Composition of the PP polymer composites formulations

Groups	PP content (wt %)	Olive pit powder content (wt %)	Almond shell powder content (wt %)
1	100	-	-
2	90	10	-
3	80	20	-
4	70	30	-
5	60	40	-
6	90	-	10
7	80	-	20
8	70	-	30
9	60	-	40

Heat deflection temperature (HDT) and Vicat softening point tests were done according to ISO 75 and ISO 307 standard with determined by CEAST 6521. HDT-Vicat test equipment. Water absorption (WA) test was carried out according to ISO 62. Three specimens for each type of composite were used in the WA test. The conditioned specimens were entirely immersed for 1 day in a container of water at 23±2 °C. At the end of immersion time, the specimens were taken out from the water and all surface water was removed with a clean dry cloth. The specimens were immediately weighed to the nearest 0.01 g. The densities of specimens were evaluated according to the test method specified in ISO 1183-1 (Method A). Three specimens for each type of composite were used in the density measurement. Static and dynamic coefficient of friction test was done according to the ISO 8295 method with Devotrans friction coefficient measurement equipment. The dimensions of the tested specimens were 80x200x4 mm and the dimensions of the sled specimens were 63x63x4 mm. Speed was selected as 100 mm/min. The static coefficient of friction μs is given by the equation,

$$\mu_S = F_S / F_P \qquad (1)$$

Where F_S is the static frictional force, expressed in Newton, F_P is the normal force exerted by the mass of the sled, expressed in Newton (=1.96 N).
The dynamic frictional force using the equation,

$$\mu_D = F_D / F_P \qquad (2)$$

Where F_D is the dynamic frictional force, expressed in Newton, F_P is the normal force exerted by the mass of the sled, expressed in Newton (=1.96 N) [10]. The wear tests were done according to the DIN 53 516 method with Devotrans DA5 abrasion test equipment. The thickness of the test specimens was 7.0 mm and diameter was 15.5 mm. Cylinder rotational speed was selected as 40 rpm and normal load (F_N) of 5, 10, 15 and 20N were used. Total sliding distance (L) of 20, 40, 60 and 80 m were used. The mass loss of the samples (Δm) was measured after the wear process, and the specific wear rates (W_S) were calculated using the following equation:

$$W_S = (\Delta m) / \rho . F_N . L \ (\text{mm}^3/\text{Nm}) \qquad (3)$$

Where Δm is the specimen's mass loss, ρ is the density of specimen, F_N is the normal load applied, and L is the total sliding distance. The friction coefficients and wear rates reported in the present study were the averages of at three measurements. The fractured surfaces of the PP polymer composites were coated to an approximate thickness of 10 nm of a gold (Au) (80%)/palladium (Pd) (20%) alloys to prevent electrical charging by Polaron SC 7620. The surfaces of the prepared samples were observed by the JEOL-JSM 5910 LV scanning electron microscopy (SEM) at an acceleration voltage of 20 kV.

3. Result and discussion:

The relationship between the filler content and the density of the polymer composites is shown in fig. 1-A. The density of the composites increased (from 0 to 40wt %) linearly with an increase weight percentage of filler. In comparison with the density of virgin PP, the density increased by 17% for the composites at a 40 wt % olive pits flour concentration. Water absorption % at different filler loading level is shown in fig. 1/B The maximum water absorption is observed at the 40 wt % olive pits flour concentration for PP. In comparison with the water absorption of virgin PP, the water absorption increased by 626% for the composites at a 40 wt % olive pits flour concentration. Fig. 1-C illustrates the effect of filler on the Vicat softening point of PP composites. The Vicat softening point increased as the olive pits and almond shell flour concentration increased from 0 to 40 wt %. The maximum Vicat softening point is observed at the 40 wt % olive pits and almond shell flour concentration.

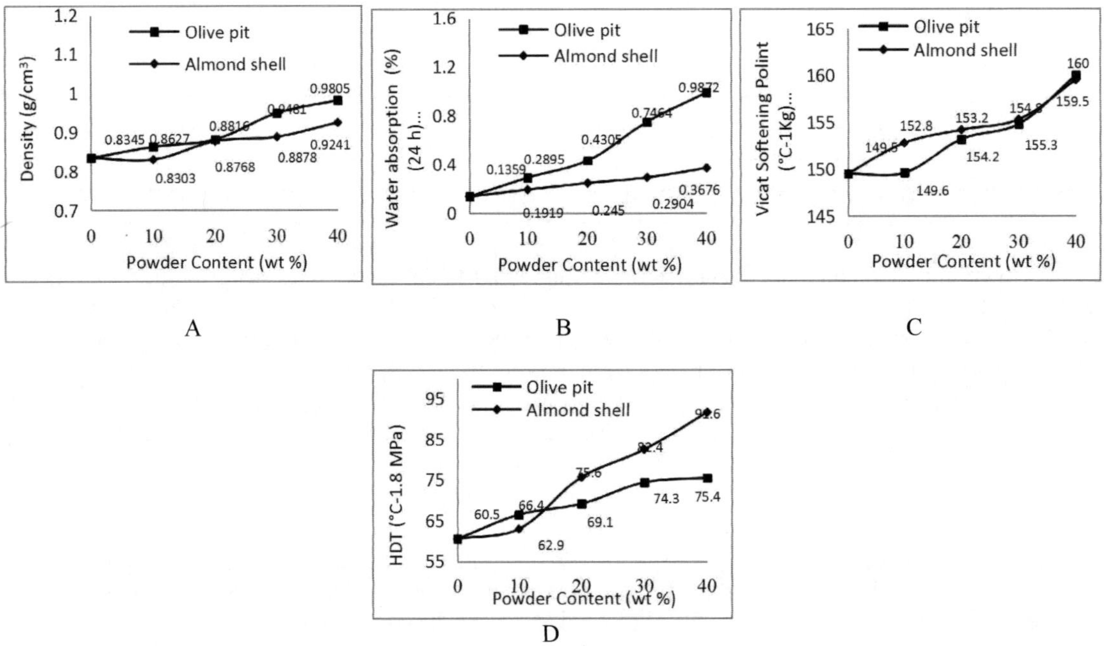

Figure 1. Properties of the PP polymer composites

The relationship between the filler content and the HDT of the polymer composites is shown in fig.1-D. The HDT of the composites increased (from 0 to 40wt %) linearly with an increase weight percentage of filler. The maximum HDT is observed at the 40 wt % almond shell flour concentration for PP. The effects of applied load, sliding distance and fillers content on the tribological behaviors of PP composites were examined. The results of the wear rate of PP/olive pit and PP/almond shell flour polymer composites at different loads and sliding distances are presented in Figure 2. It is seen that the applied load and sliding distance had a great effect on the wear rate of composites. Wear rate increased gradually with the increase of the load from 5 N to 20 N. Also, wear rate increased with the increase of the sliding distance from 20 to 40 m. It was consistent that the sliding distance and applied load affected on the wear rate of PP polymer composites. The weak bond led to the filler particles detaching from the matrix and the matrix pulling out more easily, which could increase the wear rate of the composites. Friction performance is shown in fig. 6 when speed was 100 mm/min; load separately was 1.96, 2.94, 3.92, 4.42 and 6.86 N respectively.

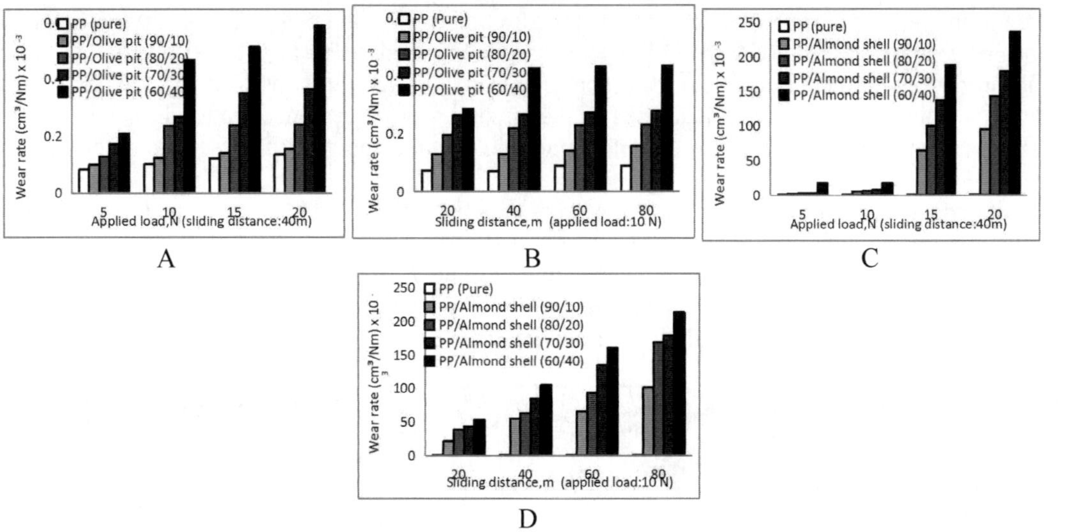

Figure 2. Wear rate of the PP/olive pit and almond shell flour polymer composites

It is seen that the load had a great effect on the static and dynamic friction coefficient of the PP/olive pits and almond shell flour composite. As the load increases, the friction coefficient of all kinds of composites increases. Fig. 3-A/B illustrates the effect of olive pit flour on the static and dynamic coefficient of friction of PP composites.

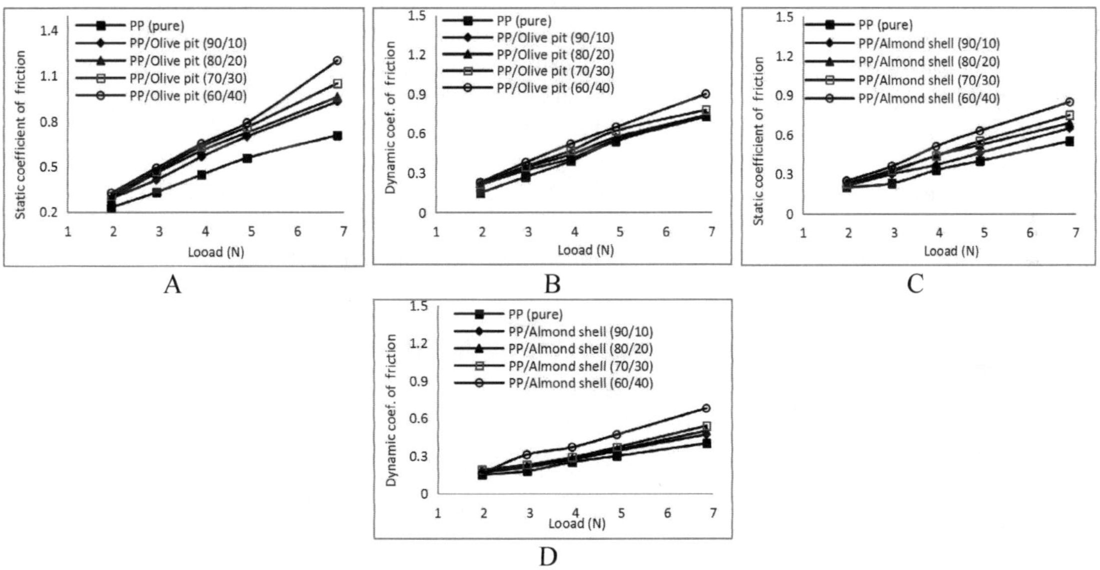

Figure 3. Coefficient of friction of the PP/olive pit and almond shell flour polymer composites

The static coefficient of friction increased as the olive pits flour concentration increased from 0 to 40 wt %. The static coefficient of friction increased as the load increased from 1.96 to 6.86 N as well. The maximum static coefficient of friction is observed at the 40 wt % olive pits flour concentration and 6.86 N load. On the other hand, the dynamic coefficient of friction increased as the olive pits flour concentration increased from 0 to 40 wt %. Also, the dynamic coefficient of friction increased as the load increased from 1.96 to 6.86 N. The maximum dynamic coefficient of friction is observed at the 40 wt % olive pits flour concentration and 6.86 N load. Fig. 3-C/D illustrates the effect of almond shell flour on the static and dynamic coefficient of friction of PP composites. The static coefficient of friction increased as the almond shell flour concentration increased from 0 to 40 wt %. The static coefficient of friction increased as the load increased from 1.96 to 6.86 N as well. On the other hand, the dynamic coefficient of friction increased as the almond shell flour concentration increased from 0 to 40 wt %. Also, the dynamic coefficient of friction increased as the load increased from 1.96 to 6.86 N.

The fractured surfaces of polymer composites were examined with SEM are presented in fig. 4. It is clear that the filler at low loading level (20%) (Fig. 4/B and D) morphology slightly differs from that of pure PP polymer (fig. 4/A). At high olive and almond shell flour loading, (%40) as shown in fig. 4C and E respectively, more filler pullout and deponding were observed. This was probably due to poor adhesion between olive and almond shell flour and polymer matrix.

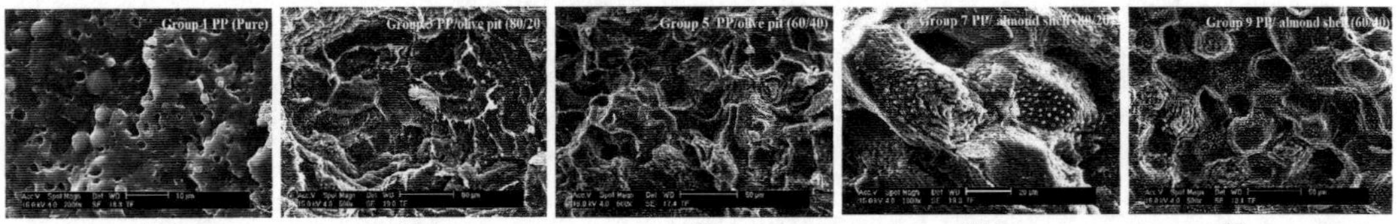

A B C D E

Figure 4. SEM micrographs of PP composite at different filler contents

4. Conclusions

The aim of this study was to investigate the effect of olive and almond shell flour at different weight fractions without coupling agent on mechanical, morphological, and water uptake properties of PP matrix composite. PP/olive and almond shell flour composite with 10,20, 30, and 40 wt. % olive and almond shell flour contents are under taken in an extrusion and compression molding processes. Substantial improvements in the mechanical properties were obtained by the addition of olive pits and almond shell flour in the PP polymer matrix. The results showed that the density, Vicat softening temperature, and HDT of composites improved with increasing powder content. On the other hand, applied load and sliding distance had a great effect on the wear rate of PP/olive pits and almond shell flour composites. Wear rate increased gradually with the increase of the load and sliding distance. Load had a great effect on the static and dynamic friction coefficient of the PP/olive pits and almond shell composite. As the load increases, the friction coefficient of all kinds of composites increases.

Acknowledgement

This work has been supported by the Scientific Research Project Program of Marmara University (No:FEN-C-YLP-091116-0497)

References

[1] Gwon, JG, Lee SY, Chun SJ, Doh GH, Kim JH, *Compos. Part A Eng.*, 2010 **41** p 1491
[2] Asim M, Abdan K, Jawaid M, Nasir M, Dashtizadeh Z, Ishak MR, Enamul HM, *Journal of Polymer Science*, 2015 **1** p 16
[3] Azwa ZN, Yousif BF, Manalo AC, Karunasena W, *Mater Des*. 2013 **47** p 424
[4] Ashori A, *Bioresour Technol.*, 2008 **99** p 4661
[5] John MJ, Thomas S, *Biofibres and biocomposites Carbohydr Polym* 2008 **71** p 343
[6] Koutsomitopoulou AF, Benezet JC, Bergeret A, Papanicolaou GC, *Powder Techn* 2014 **255** p10
[7] Banat R, Fares M M, *Int. J. of Composite Materials* 2015 **5(5)** p 133
[8] Asim M, Abdan K, Jawaid M, Nasir M, Dasgtizadeh Z, *Int. J. of Poly. Sci.* 2015 **2015** p 1
[9] Qutaiba A, "Investigation into The Modes of Damage and Failure in Natural Fiber Reinforced Epoxy Composite Materials", Ph.D. Thesis, University of East London 2011
[10] ISO 8295:1995(E) test standard: Plastics-Film and sheeting-Determination of the coefficients of friction

IOP Conf. Series: Materials Science and Engineering **204** (2017) 012016 doi:10.1088/1757-899X/204/1/012016

Effects of Waste Plastic on the Physical and Rheological Properties of Bitumen

Mohd Ezree Abdullah[1], Nurul Asyiqin Ahmad[2], Ramadhansyah Putra Jaya[2,3], Norhidayah Abdul Hassan[2], Haryati Yaacob[2],Mohd Rosli Hainin[2]

[1]Faculty of Civil and Environmental Engineering, UniversitiTun Hussein Onn Malaysia, 86400 BatuPahat, Johor, Malaysia
[2]Faculty of Civil Engineering, Department of Geotechnics and Transportation, UniversitiTeknologi Malaysia, 81310 UTM Johor Bahru, Malaysia.
[3]Faculty of Engineering, Department of Civil and Environmental Engineering, UniversitiTeknologi PETRONAS, 32610 Seri Iskandar, Perak DarulRidzuan, Malaysia

E-mail: rsyahputrajaya@gmail.com;ramadhansyah@utm.my; ezree@uthm.edu.my; hnorhidayah@utm.my; haryatiyaacob@utm.my; syiqin.pmc@gmail.com; mrosli@utm.my

Abstract. Plastic disposal is one of the major problems for developing countries like Malaysia, at the same time Malaysia needs a large network of roads for its smooth economic and social development. The limited source of bitumen needs a deep thinking to ensure fast road construction. Therefore, the use of plastic waste in road construction not only can help to protect environment but also able to help the road construction industry. The aims of this research are to study the effects of waste plastic on rheological properties of bitumen. Modified bitumen was prepared by using blending techniques. Bitumen was heated and plastic waste was slowly added. Rheological properties of bitumen were performance by penetration, softening point, viscosity and direct shear rheometer test. The results showed that when content of plastic waste increase, the penetration value, softening point and viscosity of bitumen also increase. Generally, plastic waste improves the performance of bitumen when it was added into bitumen. It can be said that the usage helps to improve the performance of the road pavement which also reduces the rutting effect.

1. Introduction

Plastics is a material that contains one or more organic polymer of large molecular weight, solid in its finished stated and it also can flow under specific state [1]. It is durable and has very slow process of degradation [2]. Plastic can be divided into two major categories which are thermoses and thermoplastics. Thermos is a condition of plastic when it is in solid form. This type of plastic is very useful in their durability and strength. Nowadays, the use of the plastic bag with several of sizes has been growing day by day [3]. This development led to an increase in the amount of waste. This hazardous waste is disposed by land filling or incineration. Waste plastic does not undergo bio-decomposition. Therefore, whether it is land filled or incinerated, it still pollutes the land and the atmosphere. However, the discovery of the binding property of plastic in its molten state which can be used in road laying has helped to well manage this waste plastic [4]. Road pavement that uses plastic waste as one of it material is called plastic road [5]. Plastic bag is non-biodegradable but most of it is

Content from this work may be used under the terms of the Creative Commons Attribution 3.0 licence. Any further distribution of this work must maintain attribution to the author(s) and the title of the work, journal citation and DOI.
Published under licence by IOP Publishing Ltd

recyclable. The recycled products are more environmentally harmful than the first time manufactured ones because every time plastic is recycled it is subject to high intensity heat. This can make it to deteriorate and lead to environmental pollution. That is why, it is necessary to determine the effective way to deal with this non-biodegradable waste [6]. The use of plastic waste in road construction can be one of the solutions [7]. This type of construction gives benefit to environment because it uses plastics that would otherwise be disposed through environmentally harmful means. Other type of methods that has been used to deal with plastic waste is by incineration [8]. However, often incinerators used are not recommended standards and guidelines. It is particularly significant to use plastic waste because it can minimize the landfilling [9]. It was translates that 1.125 tons of plastic wastes are used per km of single lane road. Based on review of literature, the objective of this research is to investigate the effect of using different percentages of plastic waste on rheological properties of bitumen. The main problems of asphalt pavement material is rutting and also cracking which is resulted from its main property of being high temperature susceptibility.

2. Material and Methods

2.1. Materials

In this study, bitumen 60/70 penetration grade was used. The properties of bitumen based on laboratory test are 65 dmm (penetration), 51 °C (softening point), 1.03 (relative density), and 600 cp (viscosity at 135°C).

2.2. Sample preparations

Modified bitumen was prepared by using blending techniques. 500 g of bitumen was heated in oven until it turns to fluid condition and plastic waste that was shredded to size of 2 mm was slowly added, while the speed of mixer was maintained at 1200-1500 rpm and the temperature was kept between 170 °C and 175 °C. The concentrations of plastic waste used were 1.5, 3.0, 4.5 and 6.0 % by weight of the bitumen. Mixing was continued for 1 hour to produce homogenous mixtures. Rheological tests such as penetration, softening point, viscosity and also direct shear test were then conducted on the prepared sample.

2.3. Bitumen test

The tests to determine rheological properties such as penetration, softening point, viscosity, and direct shear test are performed in the following study. Binders are characterized by using a number of standard physical tests such as penetration test [10] (temperature, load and time at 25 °C, 100 g and 5 sec, respectively), softening point test [11], viscosity test using Brookfield viscometer [12] (temperature ranging from 135 to 165 °C, spindle No. 27, and a rotating speed of about 20 rpm). The short-term aging test of the binder is conducted using RTFOT [13] at 163 °C for 85min; the long-term aging is performed using PAV [14] at 100 °C for 20hrs with 2.1 Mpa. The aged samples are evaluated by measuring their rheological properties. Furthermore, the bitumen rheological properties are evaluated using dynamic shear rheometer (DSR) (tests conducted by using a temperature sweep starting from 22 °C to 84 °C and the frequency of 1.59 Hz with 1mm gap [15].

3. Results and Discussions

3.1. Penetration

Fig. 1illustrates the effect of plastic waste concentration on penetration. The penetration depth decreased as the amount of plastic waste increases up to 6%. This shows that plastic waste content has a significant effect on the penetration value. The results also found that when the content of plastic waste is increased from 1.5% to 6 %, the penetration value decreases gradually from 55.3 to 24.1. It means that plastic waste has a great effect on reducing the penetration value by increasing the stiffness of plastic waste in binder. Make the binder less susceptible and will result to resist the deformation like rutting [16].

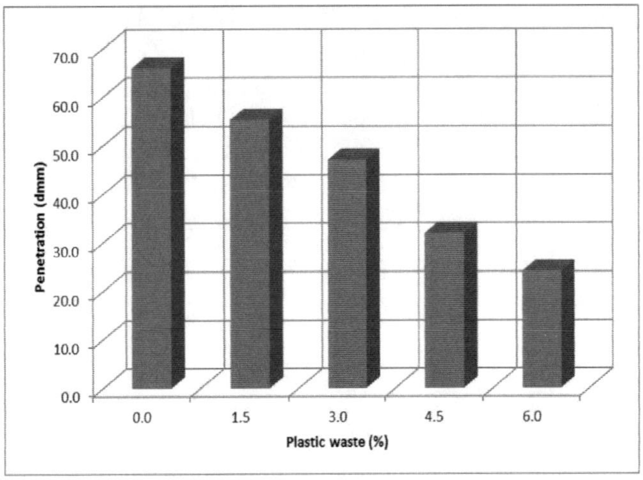

Figure 1. Penetration Values

3.2. *Softening Point*

Fig. 2 shows that the increases of softening point as the percentage of plastic waste increase. It can be seen that when plastic waste content is 1.5% by weight of bitumen, the softening point is 53 °C and when plastic waste is added up to 6%, the softening point reaches up to 56 °C. It can be said that when the softening point is increasing there will be the reduction of susceptibility at high temperature. This phenomenon shows that the resistance of the binder to the effect of heat is increased and was reduce its tendency to soften in the hot weather [17]. With the addition of plastic waste, the modified binder will be less susceptible to the changes of temperature. Therefore by using plastic waste in bituminous mix, the rate of rutting will decrease due to the increase in softening point.

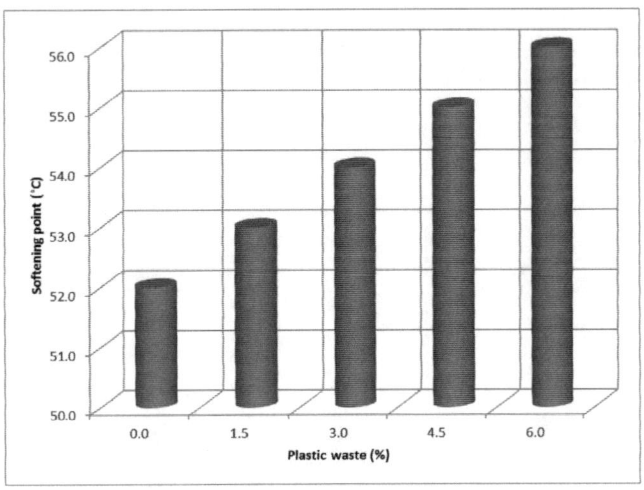

Figure 2.Softening Point Values

3.3. *Penetration Index (PI)*

The penetration index (PI) represents the quantitative measure of the response of bitumen to variation in temperature. Tables 1 summarized the value of PI for virgin and modified bitumen at different percentage of plastic waste. It can be seen that the penetration index values are inconsistent at all addition percentages. The PI value for original binder is -0.02 and decrease to -0.23 when it was modified with 1.5% of plastic waste. However, PI values were increase from -0.39 to -0.10 for 3% and 4.5% respectively. Then it decrease again from -0.10 to -1.31 at 4.5% and 6% respectively. The results show that the resistance of modified bitumen to resists the low temperature cracking and permanent deformation is inconsistent. Yaacob et al. [1] reported that the penetration index of particular bitumen can be used to predict its behavior in an application. Therefore, bitumen with high penetration numbers that can be called as soft are used for cold climates while asphalt binders with low penetration numbers that known as hard are used for warm climates [18].

Table 1. PI at different percent plastic waste

Plastic waste (%)	PI
0	-0.02
1.5	-0.23
3.0	-0.39
4.5	-0.10
6.0	-1.31

3.4. Viscosity

The viscosity of bitumen at high temperature is assumed as an important factor to have great road pavement work. This is because viscosity represents the ability of bitumen to be pumped through an asphalt plant and coating aggregates in asphalt concrete mix. The characteristics of viscosity in the asphalt binder were examined at two temperatures which are 135 °C and 165 °C. Fig. 3 illustrated the modified binder with plastic waste additive at 135 °C has greater viscosity compared to virgin bitumen. It is showed that as the percentage of plastic waste increase, the viscosity increased which means the binder is getting more viscous. Based on figure, it can be seen that the viscosity of original bitumen 60-70 is 500 cP at 135 °C which is less viscous compared to the addition of 6% plastic waste which has viscosity of 1600 cP at the same temperature. However, at temperature 165 °C the original and modified binder maintained constant 200 cP.

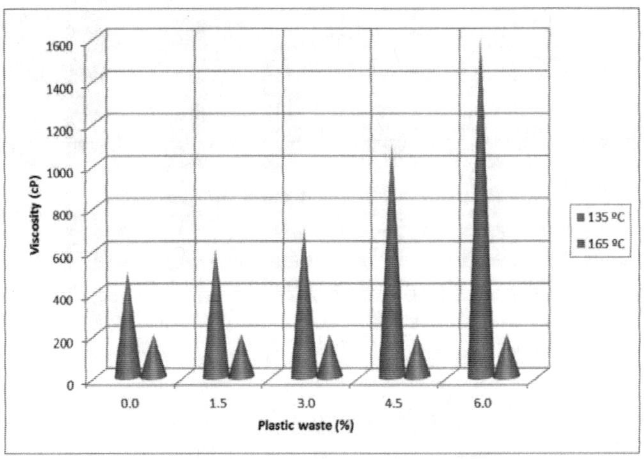

Figure 3. Viscosity value

3.5. Direct Shear Rheometer

Fig. 4 shows the G*/sin (δ) trend for original and modified bitumen at temperature ranging from 46°C to 94°C and at the constant frequency of 1.59 Hz. The results indicated that higher the percentage of plastic waste will result in higher the G*/sin (δ). It also shown that at 6% of plastic waste, the G*/sin (δ) still met the condition even the temperature reached up to 94°C. It can be said that the use of plastic waste as modifier will improve the performance of binder against permanent deformation (rutting) at high temperature. However, G*/sin (δ) at 6% plastic is lower than 4%. Indicate that a 4% of plastic waste is the optimum percentage.

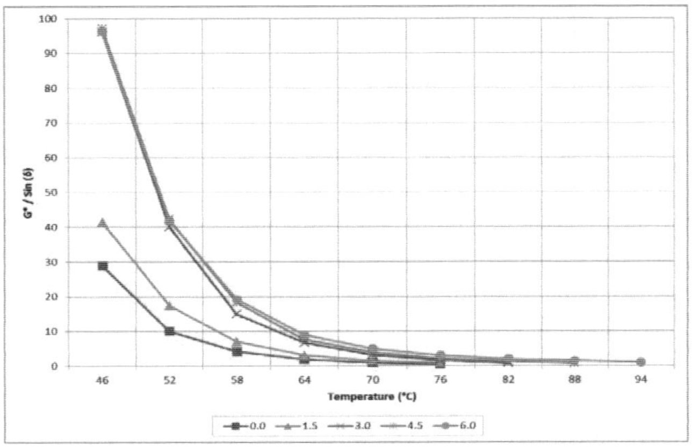

Figure 4.G*/sin (δ) trend

3.6. Penetration-Viscosity Number (PVN)

The PVN must be based on penetration at 25 °C and viscosity at 135 °C which are standard specifications for paving the asphalt. Base on Table 2, the increases of plastic waste was increasing the PVN values. The PVN value is increases from -0.36 to 0.13 which have the percentage of 0% and 6% respectively. This increases of PVN values shows the significantly of the plastic waste addition in the bitumen to improve the temperature susceptibility of bitumen. It is indicated that the addition of plastic waste in bitumen can enhance the rutting resistance of the bitumen at high temperatures.

Table 2. PVN at different percent plastic waste

Plastic waste (%)	PVN values
0	-0.36
1.5	-0.29
3.0	-0.25
4.5	-0.05
6.0	0.13

4. Conclusions

a) Additive of plastic waste at different content gives effect on the temperature susceptibility of the bitumen. As the content of plastic waste increase from 1.5% to 6%, the penetration number decreases gradually and softening point of modified bitumen increase.

b) The addition of plastic waste content increases the viscosity of the bitumen at high temperature i.e. 135 °C. High viscosity means less chances of rutting.

c) Plastic waste improves the performance of bitumen when it was added into bitumen. The higher plastic waste percentage give the higher G*/sin δ which is rutting factor. Thus, the modified bitumen able to reduce the rutting effect.
d) The usage of 6% of plastic waste is found to be the optimum percentage for modification of bitumen. This percentage could make the pavement able to resist the heavy vehicles and hot climate.

5. Acknowledgements
Research work presented in this paper was supported by Ministry of Higher Education Malaysia (MOHE), Fundamental Research Grant Scheme (research grant number 1458) and Incentive Grant Scheme for Publication (grant number U669), Universiti Tun Hussein Onn Malaysia; and Universiti Teknologi Malaysia (research grant number Q.J130000.2522. 11H58). The financial assistance for this study is highly appreciated.

References
[1] H. Yaacob, M. Ali Mughal, R.P. Jaya, M.R. Hainin, D.S. Jayanti, C.N. Che Wan, Rheological properties of styrene butadiene rubber modified bitumen binder. *J. Teknologi.***78** (2016) pp 121–126.
[2] A.I.B. Farouk, N.A. Hassan, M.Z.H. Mahmud, J. Mirza, R.P. Jaya, M.R. Hainin, H. Yaacob, N.I.M. Yusoff, Effects of mixture design variables on rubber–bitumen interaction: properties of dry mixed rubberized asphalt mixture. *Mater. Struct.* **50** (2017) pp 1-10.
[3] S. Köfteci, P. Ahmedzade, B. Kultayev, Performance evaluation of bitumen modified by various types of waste plastics. *Constr. Build. Mater.* **73** (2014) pp 592-602.
[4] M. Garcia-Morales, P. Partal, F.J. Navarro, C. Gallegos, Effect of waste polymer addition on the rheology of modified bitumen. *Fuel* **85** (2006) pp 936–943.
[5] X. Lu, U. Isacsson, Modification of road bitumens with thermoplastic polymers. *Polym. Testing* **20** (2001) pp77–86.
[6] M. Naskar, T.K. Chaki, K.S. Reddy, Effect of waste plastic as modifier on thermal stability and degradation kinetics of bitumen/waste plastics blend. *Thermochim. Acta***509** (2010) pp 128–134.
[7] C. Fang, J. Hun, S. Zhou, H. Wang, M. Zhang, Y. Zhang, Comparative study of asphalts modified by packaging waste EPS and waste PE. *Polym-Plast. Technol. Eng***50** (2011) pp 220–224.
[8] L.M.B. Costa, H.M.R.D. Silva, J.R.M. Oliveiran, S.R.M. Fernandes, Incorporation of waste plastic in asphalt binders to improve their performance in the pavement. *Int. J. Pavement Res. Technol.* **6** (2013) pp 457–464.
[9] Y. Yildirim, D. Hazlett, R. Davio, Toner-modified asphalt demonstration projects. *Resour. Conserv. Recycl.* **42** (2004) pp 295–308.
[10] ASTM D5/D5M–13. Standard Test Method for Penetration of Bituminous Materials. ASTM International, West Conshohocken, PA, USA.
[11] ASTM D36/D36M-14e1. Standard Test Method for Softening Point of Bitumen (Ring-and-Ball Apparatus). ASTM International, West Conshohocken, PA, USA.
[12] ASTM D4402/D4402M–15. Standard Test Method for Viscosity Determination of Asphalt at Elevated Temperatures Using a Rotational Viscometer. ASTM International, West Conshohocken, PA, USA.
[13] ASTM D2872-12e1. Standard Test Method for Effect of Heat and Air on a Moving Film of Asphalt (Rolling Thin-Film Oven Test). ASTM International, West Conshohocken, PA, USA.
[14] ASTM D6521–13. Standard Practice for Accelerated Aging of Asphalt Binder Using a Pressurized Aging Vessel (PAV). ASTM International, West Conshohocken, PA, USA.
[15] W.N.A.W. Azahar, R.P. Jaya, M.R. Hainin, M. Bujang, N. Ngadi, Chemical modification of waste cooking oil to improve the physical and rheological properties of asphalt binder, *Constr. Build. Mater.***126** (2016) pp 218–226.

[16] W.N.A.W. Azahar, M. Bujang, R.P. Jaya, M.R. Hainin, A. Mohamed, N. Ngadi, D.S. Jayanti, The potential of waste cooking oil as bio-asphalt for alternative binder–an overview, *J. Teknologi.* **78** (2016) pp 111–116.

[17] W.N.A.W. Azahar, M. Bujang, R.P. Jaya, M.R. Hainin, N. Ngadi, M.M.A.B. Abdullah, Performance of waste cooking oil in asphalt binder modification. *Key Eng. Mater.* **700** (2016) pp 216-226.

[18] M.R. Hainin, M.M.A. Aziz, A.M. Adnan, N.A. Hassan, R.P. Jaya, H.Y. Liu, Performance of modified asphalt binder with tire rubber powder. *J. Teknologi*73 (2015) pp 55-60.

IOP Conf. Series: Materials Science and Engineering **204** (2017) 012017 doi:10.1088/1757-899X/204/1/012017

Modelling and Optimization of Copper Electroplating Adhesion Strength

Suryanto, Farag I Haider, Mohd Hanafi Ani and M H Mahmood

Department of Manufacturing and Materials Engineering, International Islamic University Malaysia, PO Box10, Kuala Lumpur, Malaysia

E-mail:surya@iium.edu.my,fah1232003@yahoo.com,mhanafi@iium.edu.my,mahmoodfattah@yahoo.com

Abstract. In this paper, Response surface methodology (RSM) was utilized to design the experiments at the settings of $CuSO_4$ and H_2SO_4 concentrations and current densities. It also used for modelling and optimize the parameters on the adhesion strength of austenitic stainless steel substrate. The adhesion strength was investigated by the Teer ST-30 tester, and the structure of the samples investigated by using scanning electron microscopy (SEM). The modelling approach adopted in the present investigation can be used to predict the adhesion strength of the copper coatings on stainless steel substrate of electroplating parameters in ranges of $CuSO_4$ 100 to 200 g/ L, H_2SO_4 100 to 200 g / L and current density 40 to 80 mA / cm^2. The results showed that, operating condition should be controlled at 200 g/L CuSO4, 100 g/L H2SO4 and 80 mA/cm^2, to obtain the maximum adhesion strength 10N.

1. Introduction

Copper is one of the metals most extensively used in industry, either because of its intrinsic properties or as a base for further formation of metallic films. Electroplating is one of the methods most generally used to obtain metallic films of equal thickness, porosity-free construction and good adhesion [1–2]. Electroplating copper films have been widely investigated with regard to their structural characteristics, electrical properties and corrosion resistance [3-4] but less care has been yielded to their mechanical behaviour and its relation to Electroplating parameters. By controlling variables such as current density and bath concentration, a variety of pictures with different characteristics can be attained, therefore allowing to tailor the mechanical characteristics of the coatings for specific applications.

One of the most important mechanical properties of the coatings is the adhesion strength, which is defined by The ASTM as the "condition in which two surfaces are held together by either valence forces or by mechanical anchoring or by both together" [1]. (Teer ST-30) Scratch test is one of widely used, fast, and effective methods to obtain the critical loads that are related to adhesion properties of the coating [5-6]. A survey in 2001 showed that the scratch test is the most common test method used for measuring the quality of the coatings in industries [7].

In most experimental studies of concentration process, conventional methods were used to determine the influence of operational parameters. When using the conventional methods to optimize the process, one parameter is changed while others are kept at a constant level. This should be repeated for all influencing parameters, resulting in a great number of experiments[8]. Its major disadvantage is that it does not include the interactive effects among the variables studied. To overcome this problem, using a second-order model is useful in approximating a portion of the true response surface with parabolic curvature. The second-order model includes all the terms in the first-order model, plus all quadratic terms and all cross-product terms. It is usually expressed as

Content from this work may be used under the terms of the Creative Commons Attribution 3.0 licence. Any further distribution of this work must maintain attribution to the author(s) and the title of the work, journal citation and DOI.
Published under licence by IOP Publishing Ltd

IOP Conf. Series: Materials Science and Engineering **204** (2017) 012017 doi:10.1088/1757-899X/204/1/012017

Y = B0 + B1 x1 + B2 x2 + B3 x3 + B11 x21 + B22 x22 +B33 x23 +B12 x1 x2 + B13x1 x3 +B23 x2 x3 (1)

Where Y is the response (adhesion strength); xi is the variables, Bo is the constant coefficient; Bi and Bii refer to the coefficients of linear and interaction terms, respectively.

The second-order model is flexible, because it can take a variety of functional forms and approximates the response surface locally. Therefore, this model is usually a good estimation of the true response surface. There are many designs available for fitting a second-order model. The most popular one is the central composite design (CCD). This design was introduced by Box and Wilson.

The aim of this paper is to create a second-order model for the parameters ($CuSO_4$ and H_2SO_4 concentrations and current densities) and the response (adhesion strength), also to determine the optimum settings of $CuSO_4$ and H_2SO_4 concentrations and current densities that result in the maximum adhesion strength over a certain region of interest. By using RSM to design the experiments and ANOVA to analysis the data.

2. Experimental

2.1. Material

The 316L stainless steel samples and the 99.91 % pure copper were cut from 2 mm thickness sheets with 20 mm x 20 mm by EDM wire cutting machine were used as cathode and the anode respectively. All the stainless-steel samples were sanded up to 1000 SiC paper, and ultrasonically cleaned in acetone for 5 min before coated.

2.2. Experimental design

The copper thin film has been electroplated using a DC power source on a 316L stainless steel substrate with the electroplating parameters viz. $CuSO_4$ concentration (A), H_2SO_4 concentration (B), and current density (C). With $\alpha = 2$. The ranges and levels are shown in Table 1. In order to evaluate the effect of the factors on the response surface in the region of investigation, a three-factor-five-level CCD was performed.

Table 1. Experimental ranges and levels of the three independent variables used in RSM in terms of actual and coded factors

Coded value	Actual value		
	$CuSO_4$ (g/L)	H_2SO_4 (g/L)	Current density (mA/cm^2)
-1	100	100	40
0	150	150	60
1	200	200	80

The statistical software package 'Design-Expert 10 (trial version)' was used to analyse the experimental design. The total number of experiments with three factors was 20 (2k+2k+6, when k=3, where k is the number of factors). In order to control the error bar, 20 runs were performed in a random order in which there were six replications at the centre points to evaluate the purity error. Every run was done with 20 min.

2.3. Characterization

The adhesion strength of the coatings was investigated by scratch testing (Teer ST-30 tester). The Teer ST-30 is a relatively simple apparatus in which the increasing load is obtained by the inclination of the sample respect to the indenter. In the present experiments, the load was increased from 0 to 20N with an inclination degree of 1° and an indenter speed of 25 mm/ min. The morphology of the copper thin film has been studied using Scanning Electron Microscope (SEM, JSM-6380, JEOL).

IOP Conf. Series: Materials Science and Engineering **204** (2017) 012017 doi:10.1088/1757-899X/204/1/012017

3. Resultsand discussion

3.1. Results

Figure 1, shows the morphological details of the copper coating by SEM. The adhesion behaviour of the coating was assessed by analysing the representative images of the grooves left after the scratch tests shown in figure. 2. In general, the adhesion behaviour was better for the C than A, B.

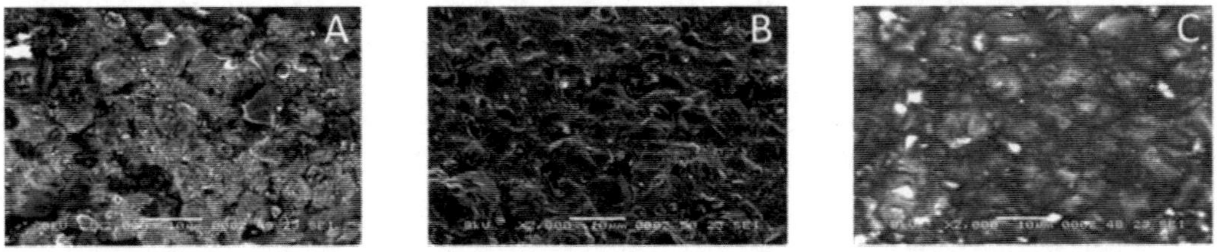

Figure 1.SEM morphology of coatings produced in $CuSO_4$, H_2SO_4 concentrations and current densities respectively as A) 200 g/L – 106 g/L – 80 mA/cm^2. B) 200 g/L – 100 g/L – 57 mA/cm^2. C) 200 g/L – 100 g/L – 80 mA/cm^2.

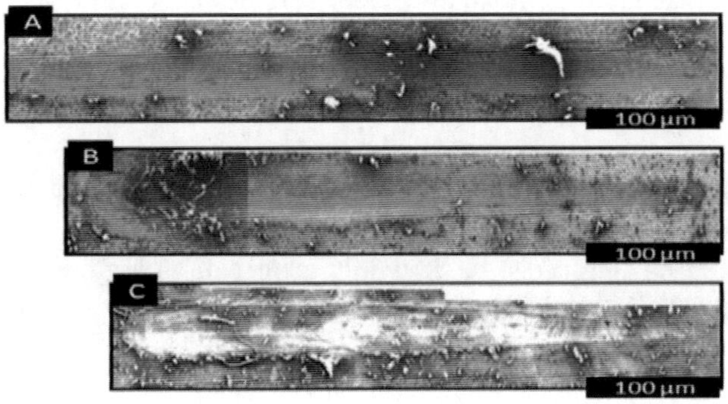

Figure 2. The scratch adhesion tests on coating obtained in $CuSO_4$, H_2SO_4 concentrations and current densities respectively as

A) 200 g/L – 106 g/L – 80 mA/cm^2. B) 200 g/L – 100 g/L – 57 mA/cm^2. C) 200 g/L – 100 g/L – 80 mA/cm^2

3.2. Statistical Analysis

The analysis of variance (ANOVA) was applied to study the effect of the input parameters on the adhesion strength. Table 2, gives the statistics for the model summary. The R-Squared of 0.9852 is closed to 1, and the Pred R-Squared of 0.9350 is in reasonable agreement with the Adj R-Squared of 0.9733; i.e. the difference is less than 0.2. It reveals that the quadratic model is the best appropriate model. So, for further analysis this model was used. Table 3, gives the Estimated Regression Coefficients of the adhesion strength for uncoded units. The value ''p'' for the model of <0.0001 is less than 0.05 which appear that the model terms are significant except C-current density with P-value more than 0.05 which is indicates that the current density as term C is less significant, but the terms AC and BC have a significant effect on the response.

Table 2. Model summary statistics.

R-Squared	Adj R-Squared	Pred R-Squared	Adeq Precision
0.9852	0.9733	0.9350	27.152

Table 3.Estimated regression coefficients of the adhesion strength.

Source	Sum of Squares	df	Mean Square	F-value	P-value	
Model	11.80	8	1.47	82.93	< 0.0001	significant
A-$CuSO_4$	4.04	1	4.04	226.97	< 0.0001	

B-H_2SO_4	3.97	1	3.97	223.32	< 0.0001	
C-Current	2.624E-003	1	2.624E-003	0.15	0.7089	
AB	0.26	1	0.26	14.68	0.0033	
AC	0.55	1	0.55	31.13	0.0002	
BC	0.46	1	0.46	25.94	0.0005	
A2	1.44	1	1.44	81.11	< 0.0001	
B2	1.78	1	1.78	99.85	< 0.0001	
Residual	0.18	10	0.018			
Lack of Fit	0.12	5	0.024	2.04	0.2268	Not significant
Pure Error	0.059	5	0.012			
Cor Total	11.97	18				

The effect of $CuSO_4$ and H_2SO_4 concentrations and the Current densities were investigated in order to serve as a tool for predicting final film adhesion strength. The quadratic equation for predicting the optimum point was obtained according to the Central Composite Design (CCD) and input variable, and then the empirical relationship between the response and the independent variable in the uncoded units was presented on the basis of the experimental results as follows:

Sqrt (Adhesion strength) = 3.66237-0.023012 (CuSO4) -0.012915 (H2SO4) -4.64939E-03(Current density) -8.31768E-05 (CuSO4) (H2SO4) +3.02834E-04 (CuSO4) (Current density) – 2.76441E-04 (H2SO4) (Current density) +9.38342E-05 $(CuSO4)^2$+1.04111E-04 $(H2SO4)^2$(2)

This developed mathematical model can be used to predict the adhesion strength of copper electroplating. Also, accuracy of the model was determined by conducting conformity test runs using the same coating process. In this procedure, the process variables were assigned random values in order to carry out the validation test runs and the responses were measured and recorded as shown in Table 4. The results show that the model is accurate.

Table 4. Comparison of predicted and actual values of adhesion strength

No.	$CuSO_4$ (g/L)	H_2SO_4 (g/L)	Current Density (mA/cm^2)	Predicted Adhesion strength (N)	Actual Adhesion strength (N)	Error %
1	200	119	80	6.83	7	2.5
2	200	100	57	6.30	6	4.8
3	200	100	80	10.01	9.50	5.1

3.3. Optimization

Based on analysing the equation. (2). With statistical experimental method, the optimal condition for adhesion strength of copper electroplating should be obtained for coating application. The adhesion strength of copper film, 10 N, can be obtained if the operation condition controlled at 200 g/L $CuSO_4$, 100 g/L H_2SO_4 and 80 mA/cm^2. As figure. 3. Validation experiments were performed and the results showed that the difference between the predicted values and the measured values is within 5.1%. Statistical optimization method overcomes the limitations of classic empirical methods and is proved to be a powerful tool for the optimization of copper film deposition.

4. Conclusions

The developed mathematical model equation (2) is accurate to predict the adhesion strength value at the ranges of CuSO4 100 to 200 g/L, H2SO4 100 to 200 g/L and Current density 40 to 80 mA/ cm2. And it is found that, the maximum adhesion strength 10 N, could be obtained at operating condition 200 g/L CuSO4, 100 g/L H2SO4 and 80 mA/cm2.

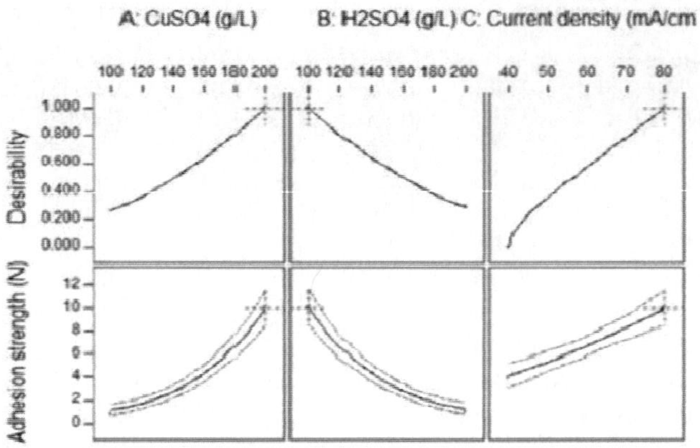

Figure 3.Optimization of the adhesion strength

References

[1] M. Schlesinger M 2010Paunovic, Modern Electroplating(*Wiley New York* **5thedition**) p.33.

[2] Yamasaki T P Schlopmacher K Ehrlich and YOgino1998 Formation of amorphous electrodeposited Ni-W alloys and their nanocrystallization(*Nanostructured Materials* **10-3**)pp 375-388.

[3] Wang Chen Jipu Lei Cameron Bjelkevig Sergei Rudenja Noel Magtoto and Jeff Kelber2003 Electrodeposition of adherent copper film on unmodified tungsten(*Thin solid films***445-1**)pp 72-79.

[4] Seah C H S Mridha and L H Chan 2001DC/pulse plating of copper for trench/via filling *(Journal of Materials Processing Technology* **114-3**)pp 233-239.

[5] Melo-Máximo L O Salas D MeloMáximo JOseguera Victor Manuel Lopez Hirata R D Torres C M Lepienski and R M de Souza 2013 Performance of Cr oxide coatings on 304 steel against metal dustingSurface and Coatings(*Technology* **237**) pp 39-50.

[6] Uribe E O Salas D MeloMáximo JOseguera C M Lepienski R D Torres and R M Souza2015 Development of Al oxide PVD coatings against metal dusting (*Surface Engineering* **31-2**)pp 114-122.

[7] Gonczy Stephen T and Nicholas Randall2005 An ASTM standard for quantitative scratch adhesion testing of thin, hard ceramic coatings (*International Journal of Applied Ceramic Technology***2 – 5**) pp 422-428.

[8] Khataee A R MFathinia S Aberand M Zarei2010 Optimization of photocatalytic treatment of dye solution on supported TiO_2 nanoparticles by central composite design: intermediates identification(*Journal of hazardous materials***181-1**) pp 886-897.

[9] Raymond H Myers Douglas C Montgomery and Christine M Anderson 2009 Response surface methodology process and product optimization using design experiments (*Wiley* **3rd edition**) pp 220.

Characterization of the microfibrillated cellulose from water hyacinth pulp after alkali treatment and wet blending

M Asrofi[1], H Abral[2], A Kasim[3] and A Pratoto[2]

[1] Doctoral Student of Agricultural Sciences, Faculty of Graduate Studies, Andalas University, 25163 Padang, Indonesia
[2] Department of Mechanical Engineering, Andalas University, 25163 Padang, Indonesia
[3] Department of Agriculture Technology, Andalas University, 25163 Padang, Indonesia

E-mail: abral@ft.unand.ac.id

Abstract. In this study, the effect of mechanical treatment on the characterization of microfibrillated cellulose (MFC) from water hyacinth pulp was carried out. Chemical and mechanical treatments were used in this experiments. Chemical treatment such as cooking fibers into the high-pressure reactor.While, the mechanical treatment that provides shear stress on the fibers by means of the wet blending process. The experiments were conducted at the time of wet blending namely 15, 20, and 25 min respectively. The final products were characterized by using scanning electron microscopy (SEM), X-ray diffraction (XRD), and tensile test. SEM observation was carried out on the surface of the film. XRD shows that the crystallinity index of MFC with wet blending 15, 20, and 25 min are 56.24, 78.41, and 85.97% respectively. The maximum value of tensile strength was 23.21 MPa at 25 min of wet blending.

1. Introduction

The last two decades, the potential of natural fibers reinforced biocomposites become interest topics to replace conventional composite materials due to its environmentally, friendly, and availability. The availability of natural fibers and their use in the micro to nano size fibers as a reinforcing agent becomes a main focus of researchers [1].

The natural fibers usually consist of cellulose, hemicelluloses, and lignin. Cellulose is a fibril component of the plant cell. Chemically, cellulose is a linear polymer of residual impurities (1 => 4) - linked β-D-glucopyranose. In nature, cellulose chain is packaged in the form of microfibril which is stabilized by intermolecular and intramolecular hydrogen bonds [2].

Cellulose with a new morphology was developed by Turbak et al. (1983) which is known as micro fibril cellulose (MFC) [3]. The fabrication of MFC pulp through a mechanical process able to produce cellulose with large surface area. The advantages have been used for any aspect such as additives in food, paint, cosmetics, and medical product. The MFC can also be used as reinforcement to the thermosetting polymers and thermoplastic polymers [1,4].

The most fiber in form of pulp has a degree of polymerization in range 600-1500. Cellulose can be found in cell walls and woody parts of plants. Cellulose has an important role in determining the fiber properties in the fabrication of paper. Making pulp fibers with high cellulose content have to observe the properties of cellulose content especially on degree polymerization of the cellulose molecule [5,6].

Content from this work may be used under the terms of the Creative Commons Attribution 3.0 licence. Any further distribution of this work must maintain attribution to the author(s) and the title of the work, journal citation and DOI.
Published under licence by IOP Publishing Ltd

IOP Conf. Series: Materials Science and Engineering **204** (2017) 012018 doi:10.1088/1757-899X/204/1/012018

The shape of pulp fiber is usually produced by using mechanical methods. However, the method for the destruction of natural fiber can damage the microfibril structure by reducing molar mass and degree crystallinity of natural fibers [7].

Takashi Taniguchi and Yano report on how to produce the microfibrils by the simple mechanical method by means of the super grinding method. This method works by giving shear stress to the longitudinal sample of fiber. The size of fiber was produced by this method is in the range of 20-90 nm diameter [8,9].

Moreover, Herrick, Nakagito, and Henrickson produce pulp fibers by passing a suspension of cellulose pulp fibers through the mechanical homogenizer. This process can increase the degree of fibrillation [4,10,11].

However, according to our best knowledge, no studies have reported on the effect of wet blending time on the characteristics of microfibrillated cellulose (MFC) from water hyacinth pulp. In this study, a tensile test was carried out to evaluate the tensile strength. The surface morphology of the MFC was investigated by scanning electron microscopy. XRD experiment was performed to know the index crystallinity.

2. Materials and Methods

2.1. Materials

The water hyacinth plants were obtained from Payakumbuh, Indonesia. The content of lignin and cellulose were 5% and 43% respectively. Sodium Hydroxide (NaOH) technical grade was obtained from Bratachem Company, Padang, Indonesia. Distilled water was supplied from the Faculty of Mathematics and Natural Science, Andalas University.

2.2. Preparation of MFC

The roots and leaves of the water hyacinth plant were separated from the rod and cleaned from dirt. Then, the rods were dried in plastic home UV (ultraviolet) for 5 days and cut along ± 1 cm before being alkalization processed into the high-pressure reactor. The water hyacinth rods, distilled water, and sodium hydroxide (NaOH) 25% for alkaline treatment were added into the high-pressure reactor with a ratio 1 kilogram, 10 liters, and 250 grams respectively, then stir the mixture. After that, they were heated at temperature and pressure of 130°C and 2 bars respectively for 6 h. Finally, we produce soft MFC with pH of about 12-13. Then, the MFC was neutralized with distilled water until pH 7. After alkalization treatment into the high-pressure reactor, the MFC was blended in wet condition at 25000 rpm using ice blender under the Philips brand. Varied time of wet blending treatment was 15, 20, and 25 min respectively. After that, they were cast in a pulp paper into screen printing 30 x 40 cm with sieve mesh of 62T. And then dried in the sun for one day before being characterized.

2.3. Surface morphology analysis

The surface morphology of the sample films was investigated by using a Scanning Electron Microscope (SEM) ZEISS EVO MA10 with an acceleration voltage of 10 kV. The surface of the sample was coated with Palladium-Gold (Pd-Au) using an Emitech SC7620 Sputter Coater machine and then observed.

2.4. Crystallinity index analysis

Wide angle of X-Ray Diffraction (XRD) was measured using a PANalyticalXpert PRO. Cu Kα radiation (λ=0.1542 nm) operating at 40 kV and 35 mA. The range of scattering angle was recorded at (2θ = 10-30°) (θ is the Bragg's angle). Crystallinity index was calculated by using Segal's method [12]:

$$I_{cr}=\left(\frac{I_{200}-I_{am}}{I_{200}}\right)\text{ x }100\% \tag{1}$$

Equation (1) indicates that I_{200} is the maximum intensity of the diffraction peak 200 ($2\theta = 22.6^{o}$) and I_{am} is the intensity of diffraction ($2\theta = 18^{o}$). I_{200} was indicated crystallinity and amorphous regions. Meanwhile, I_{am} was indicated only amorphous regions.

2.5. Tensile strength

There were three specimens were tested for every sample. The test was carried out after 2 days of releasing samples from the screen printing mold. The standard of the sample for testing was ASTM D638. Tensile tests were conducted with an Equipment Com-Ten testing machine of 95T Series. Both speed and temperature during testing were 3 mm/min and room temperature, respectively.

3. Results and Discussion

3.1. Surface morphology studies

SEM was used to know the morphological change of MFC by wet blending treatment at 15, 20, and 25 min, respectively. SEM images of MFC from water hyacinth pulp is presented in figure 1(a-c) with the same magnification (x25000).

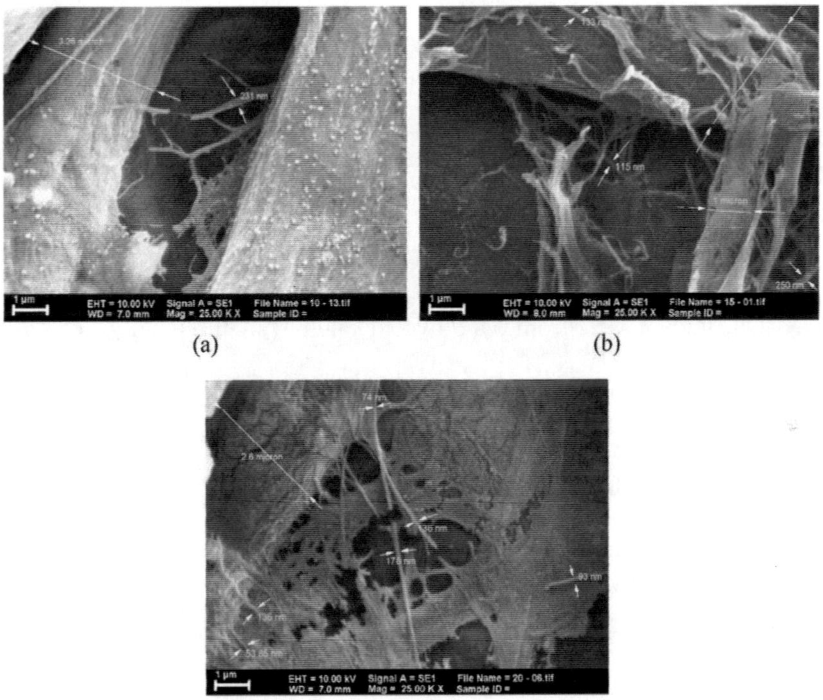

(a) (b) (c)

Figure 1. SEM on surface morphology of MFC from water hyacinth pulp: (a) wet blending time for 15 min; (b) wet blending time for 20 min; (c) wet blending time for 25 min.

As shown in figure1, MFC shows diameter around 3 μm on the surface of the sample with blending time for 15 min. Those fibers, showed by SEM images for MFC samples figure 1(b-c) are disintegrated from micro to nano-size after 20 and 25 min of wet blending were performed, respectively. It was indicated, the degree of fibrillation of MFC from water hyacinth pulp was increased with the increased time of wet blending treatments. However, the amount of MFC from water hyacinth pulp in nano-size is minimum. A substantial part of fibrils is around 50 – 100 nm in diameter. This phenomenon similar with previous report [1,4,9].

3.2. Cristallinity index studies

Figure 2 shows the diffraction pattern of the MFC from water hyacinth pulp with wet blending treatment for 15, 20, and 25 min showed two sharp peaks at ($2\theta = 16^o$ and 22.6^o) respectively. On the XRD peak of ($2\theta = 22.6^o$) and ($2\theta = 16^o$) correspond with crystallinity and amorphous region respectively. This phenomenon was indicated the crystal structure of cellulose type I as reported by Sunil et.al. and Ahmad et. al. due to the formation of inter and intramolecular H-bonding in cellulose by the hydroxyl groups [13,14].

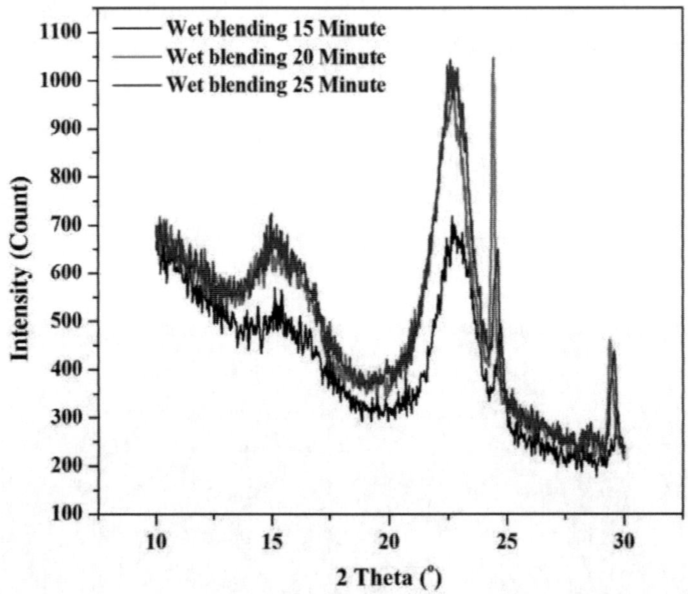

Figure 2. XRD pattern of MFC from water hyacinth pulp with different time of wet blending

The crystallinity value was calculated by using Segal's method. The estimated crystallinity value of MFC with wet blending time for 15, 20, and 25 min were 56.24, 78.41, and 85.97% respectively. It was indicated, the role of the alkali and wet blending treatments were very significant to improve the crystallinity of cellulose fibers due to the removal of the amorphous region such as hemicellulose and lignin contents during treatments. This result is in agreement with the Sunil et. al.'s report that chemical treatment such as alkalization, acid hydrolysis, and bleaching able to remove amorphous region and present a crystallinity component in the samples [13]. However, the higher crystallinity of the cellulose fibers can also increase the tensile strength as reported by Alemdar et. al. and Janoobi et. al. [2,12].

3.3. Tensile strength studies

As shown in figure 3, the tensile strength of the MFC from water hyacinth pulp in membrane films with different time of wet blending (15, 20, and 25 min). The tensile strength of the MFC with wet blending time for 15 min was about 18.41 MPa. The tensile strength slightly increased with the increased time of wet blending up to 23.21 MPa at 25 min. The MFC did improve the tensile strength due to the formation of inter and intramolecular H-bonding by the hydroxyl groups as reported by Janoobi et al and Sunil et al. [12,13]. The results are in agreement with the SEM and XRD observations that the higher degree of fibrillation and crystallinity of fibers can increase the tensile strength.

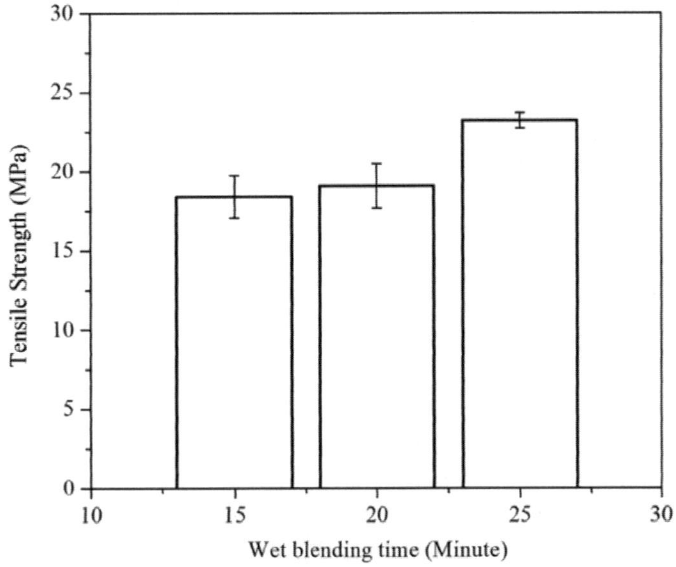

Figure 3. The tensile strength of MFC from water hyacinth pulp with different time of wet blending

4. Conclusion

The fabrication of MFC can be done with chemical and mechanical methods. A chemical method such as cooking fibers into the high-pressure reactor with a 25% sodium hydroxide solution at temperature and pressure was about 130°C and 2 bars respectively then proceed by mechanical treatment such as wet blending treatments with ice blender. The tensile strength slightly increased with the increased time of wet blending. The maximum value of tensile strength was 23.21 MPa at 25 min of wet blending time. XRD shows that the crystallinity index of cellulose MFC with wet blending time for 15, 20, and 25 min are 56.24, 78.41, and 85.97% respectively. It was indicated that the alkali treatment and wet blending were successful to removal of the amorphous region such as hemicellulose and lignin contents in WH fibers. These results are in agreement with the tensile strength observations that higher crystallinity of fibers can increase the tensile strength as reported by Janoobi. SEM micrograph on surface morphology shows a reduction size of fiber in diameter with 15 to 25 min of wet blending time was from ≥ 1 μm to 50-100 nm respectively. In this study, the bleaching and acid hydrolysis processes were not performed. Therefore, this study need to be studied further in terms of its chemical treatment, especially in the alkalization, acid hydrolysis, and bleaching processes. However, the diameter size, crystallinity, and mechanical properties of the natural fibers were very significantly determined by chemical and mechanical treatments.

Acknowledgement

This research was funded by Directorate General of Higher Education Ministry of National Education, Indonesia, with project name The Research of Master Program Leading to Doctoral Degree for Excellent Students in the year of 2016.

References

[1] Zhang J., Song H., Lin Lin., Zhuang J., Pang C., and Liu S. Microfibrillated cellulose from bamboo pulp and its properties. *Biomass and Bioenergy*, **39**, pp. 78–83, 2010.
[2] Alemdar A. and Sain M. Isolation and characterization of nanofibers from agricultural residues-

wheat straw and soy hulls. *Bioresource Technology*, **99**, pp. 1664–1671, 2008.

[3] TurbakA.F., Snyder F. W., and Sandberg K. R. Microfibrillated Cellulose, a New Cellulose Product: Properties, Uses, and Commercial Potential. *J. Appl. Polym. Sci, Appl. Polym. Symp*, **37**, 1983.

[4] Nakagaito A. N. and Yano H. The effect of morphological changes from pulp fiber towards nano-scale fibrillated cellulose on the mechanical properties of high-strength plant fiber based composites. *Applied Physics A*, **552**, pp. 547–552, 2004.

[5] Xu G. Wet Strength Improvement of Paper Via Crosslinking of Cellulose Using Polymeric Carboxylic Acids and Aldehydes. University of Georgia, 2001.

[6] Plackett D. Microfibrillated cellulose and new nanocomposite materials : a review. *Cellulose*, pp. 459–494, 2010

[7] Li J., Wei X., Wang Q., Chen J., Chang G., Kong L., and Su J. Homogeneous isolation of nanocellulose from sugarcane bagasse by high pressure homogenization.*Carbohydrate Polymers*, **90**, no. 4, pp. 1609–1613, 2012.

[8] Taniguchi T. and Okamura K. New films produced from microfibrillated natural fibres.*Polymer International*, **47**, no. 3, pp. 291–294, 1998.

[9] Uetani K. and Yano H. Nanofibrillation of Wood Pulp Using a High-Speed Blender. *Biomacromolecules*, **12**, no. 2, pp. 348–353, 2011.

[10] Henriksson M., Henriksson G., Berglund L. A., and Lindstro T. An environmentally friendly method for enzyme-assisted preparation of microfibrillated cellulose (MFC) nanofibers. *European Polymer Journal*, **43**, pp. 3434–3441, 2007.

[11] Lee J. J. S., Endo T., and Kim N. Characteristics of microfibrillated cellulosic fibers and paper sheets from Korean white pine.*Wood Sci Technol*, pp. 3–11, 2013.

[12] Janoobi M., Harun J., Shakeri A., Misra M., and Oksman K. Chemical composition, crystallinity, and thermal degradation of bleached and unbleached kenaf bast. *Bioresources*, **4**, pp. 626–639, 2009.

[13] George J. C. C. S. N. and Narayanankutty S. K. Isolation and characterization of cellulose nanofibrils from arecanut husk fibre.*Carbohydrate Polymers*, **142**, pp. 158–166, 2016.

[14] Sheltami R. M., Abdullah I., Ahmad I., Dufresne A., and Kargarzadeh H. Extraction of cellulose nanocrystals from mengkuang leaves (Pandanus tectorius). *Carbohydrate Polymers*, **88**, no. 2, pp. 772–779, 2012.

IOP Conf. Series: Materials Science and Engineering **204** (2017) 012019 doi:10.1088/1757-899X/204/1/012019

Optimization of Temperatures Heating Melt and Annealing Soft Magnetic Alloys

Vladimir Tsepelev[1] and Yuri Starodubtsev[1,2]

[1]Boris Yeltzin Ural Federal University, 19 Mira str., Yekaterinburg, 620002, Russia
[2]Gammamet Research and Production Enterprise, 28 Kirov str., Yekaterinburg, 620028, Russia

E-mail: v.s.tsepelev@urfu.ru, yunstar@mail.ru

Abstract. Taking into account the concept of the quasi-chemical model of the liquid micro-non-uniform composition and the research made on the physical properties of the Fe-based melts being crystallized, the unique technology of the melt time-temperature treatment has been developed. Amorphous ribbons produced using this technology require optimal annealing temperatures to be specifically selected. Temperature dependences of the kinematic viscosity of a multicomponent $Fe_{72.5}Cu_1Nb_2Mo_{1.5}Si_{14}B_9$ melt have been studied. A critical temperature is detected above which the activation energy of viscous flow of the melt changes. Upon cooling the overheated melt, the temperature curves of the kinematic viscosity become linear within the given coordinates. In amorphous ribbon produced in the mode with overheating the melt above the critical temperature, the enthalpy of crystallization grows, the following heat treatment results in an increase in magnetic permeability.

1. Introduction

The concept of the quasi-chemical model of the liquid micro-non-uniform composition is being developed in our Research Center [1]. According to it, the metal melt consists of space areas (groups, cybotaxis or clusters) within which the atom arrangement is characterized by certain ordering – short-range order. Due to moderately intense particles' heat movement, the clusters have no clear-cut boundaries: the atoms' ordered arrangement is being continuously replaced by another one, moving away from the cluster's core. For the same reason, the time of the given cluster's existence is limited and depends upon the chemical bonds' energy. At the same temperature, it is possible for two or more cluster ordering types to co-exist.

Soft magnetic nanocrystalline materials are distinguished by low coercivity and high magnetic permeability magnitudes, which are provided by nanocrystalline, structure [2]. The fabrication of these materials includes three stages. At the first stage, an alloy of a given composition is melted. Then, using ultrarapid quenching, amorphous metallic ribbon about 25 μm in thickness is produced. At the third stage, in the course of heat treatment, in the ribbon there forms an optimum nanocrystalline structure.

The effect of the temperature of preparation of the melt prior to casting, which is aimed at creation in the iron-based alloys of amorphous structure, was investigated in a number of works [3–6]. It was shown that, with increasing temperature of the melt prior to casting, the ribbon becomes more plastic, its electrical resistance and crystallization temperature grow, and its density lowers. On the other hand, the increased temperature of the melt deteriorates the quality of the ribbon surface and efficiency of its production. Therefore, it is important to find an optimum mode of preparation of the melt prior to casting. As a characteristic point for a multicomponent melt, a critical temperature T_k can be specified

Content from this work may be used under the terms of the Creative Commons Attribution 3.0 licence. Any further distribution of this work must maintain attribution to the author(s) and the title of the work, journal citation and DOI.
Published under licence by IOP Publishing Ltd

IOP Conf. Series: Materials Science and Engineering **204** (2017) 012019 doi:10.1088/1757-899X/204/1/012019

[6], above which the properties of the melt irreversibly change upon cooling. Irreversibility manifests itself in a hysteresis of kinematic viscosity, surface tension, and other properties of the melt. The hysteresis can arise as a result of breaking of interatomic bonds which control the short-range order in the melt.

In the paper, the results of measurements of the viscosity of the $Fe_{72.5}Cu_1Nb_2Mo_{1.5}Si_{14}B_9$ melt are presented. The parameters of process of crystallization and magnetic properties of ribbons produced upon different modes of preparation of the melt prior to casting are investigated.

2. Experimental

The main method used in this work – a torsional oscillatory method of measurement of kinematic viscosity of metal melt. The main measured size – logarithmic decrement of attenuation of torsional fluctuations of a crucible with metal melt. In a consequence logarithmic decrement of attenuation is recalculated to kinematic viscosity on Shvidkovsky's equations [7].

Distinctive features of the used experimental plant:

Measurement of logarithmic decrement is completely automatic [8]. It excludes influence of "a human factor" on results of measurements. Besides, the automatic mode allows to receive the maximum quantity of data for a unit of time. One measurement of logarithmic decrement of attenuation of fluctuations takes about 40 seconds.

The original system of an initial swing of suspended system is used. Use of this system allows to carry out a fast swing and to avoid emergence of cross fluctuations of suspended system [9].

The special software is used. It allows to use the steps counter of the computer as the counter for measurements of time. It provides high precision of measurements.

Experiments were made in the atmosphere of helium. The internal volume of experimental installation was pumped out to vacuum of $3\cdot10^{-3}$ mm Hg before filling of the camera with helium. The mass of a sample is about 25 grams. Diameter of a sample is about 13 millimeters. Height of a sample is about 20 millimeters. Crucibles from beryllium oxide are used. The period of fluctuations of suspended system is about 4 seconds. Heating is carried out by means of the heater made of molybdenum. Regulation of heating was carried out by means of the simistor block by a pulse-phase method. Temperature was controlled with the help of tungsten-rhenium thermocouple. Regulation of temperature was carried out automatically with use of proportional integrated differential algorithm.

$Fe_{72,5}Cu_1Nb_2Mo_{1,5}Si_{14}B_9$ alloys were melt in a vacuum induction furnace. A 25 μm thick and 10 mm wide ribbon with the amorphous structure was produced by the planar flow casting process. The ribbon was wound up onto ring-type cores with the 32 mm outer diameter and 20 mm inner diameter. The cores were subjected to annealing at $T_a = 300 – 580$ °C in the air for $t = 1$ hour, following the crystallization peak.

3. Results and Discussion

In the general form, the time-temperature treatment cycle is graphically represented in Fig. 1 in temperature-time coordinates. The most significant part is the establishment of the technological parameters, vis., the maximum temperature of heating T_{treat} exposure time of the melt at this temperature (t_{treat}), the temperature of the metal before exiting from the furnace t_{exit}. One variant of the procedure of determining t_{treat} is to measure the physical properties of the melt during its holding at T_{treat}, which is somewhat higher than T_k. The relaxation time of the structure and stabilization of the properties can be recommended as the value of t_{treat}. Stirring the metal proves to be effective. It enables one to decrease the holding time at the critical temperature.

The development of specific TTT technologies for a more efficient application of their potentials stipulates the changes possible in the order of alloying components and deoxidizers added, the exposure temperature of the metal, correction of the heat-treatment conditions of the parts, etc.

The typical temperature dependences of kinematic viscosity ν, electrical-resistance ρ, magnetic susceptibility χ, surface tension σ, and density d of the liquid amorphous alloys are shown in Fig. 2. It is found that the critical temperatures termed T_k are determined on basis of such data. Heating above T_k ensures a more homogenous melt state and, as a rule, causes the formation of the physical properties hysteresis. Overcooling ΔT is measured for this equilibrium melt. The presence of hysteresis is indicative for the differences between the initial and final melt states at T_k in the course of melting. Sometimes, the melt may be heated only up to the anomaly temperature T_{an}. The melt is held at the maximum heating temperature for a particular time during the experiment, investigating the time dependence of the properties changes at this temperature.

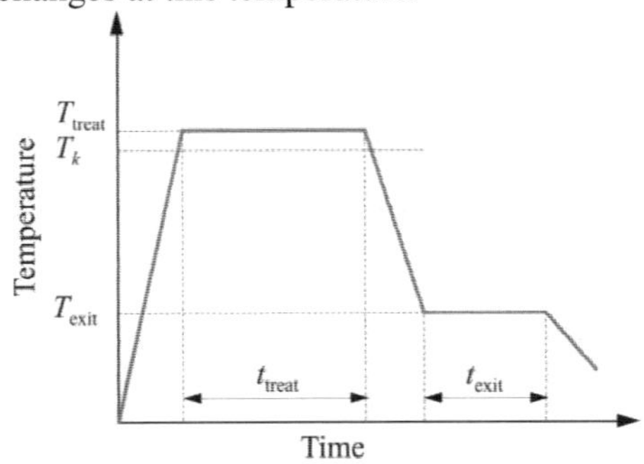

Figure 1. Thermogram of the melting of a commercial metal depicting the temperature-time treatment of the melt.

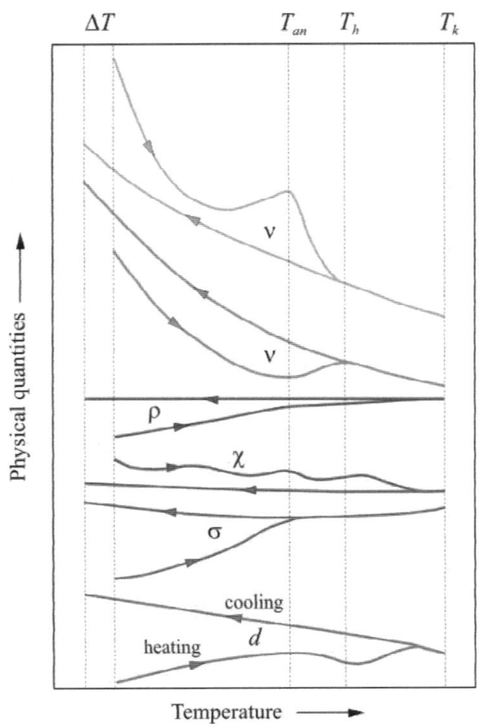

Figure 2. The typical polytherms of physical properties of amorphous alloys: ΔT is overcooling; T_{an} is the temperature of anomalies; T_h is the temperature of hysteresis; T_k is the critical temperature.

In Fig. 3, the experimental results of the viscosity measurements are presented in the form of dependence of the inverse kinematic viscosity of the melt, 1/ν, on the inverse absolute temperature, $1/T$. In the case of heating (Fig. 3a), the curve is divided into two linear portions. The change in the slope takes place at a temperature of 1490 °C. This value coincides with the critical temperature T_k obtained earlier using the Arrhenius law. In the course of cooling (Fig. 3b), the dependence of the inverse kinematic viscosity 1/ν on the melt inverse absolute temperature $1/T$ behaves linearly with a constant slope over the whole temperature range.

IOP Conf. Series: Materials Science and Engineering **204** (2017) 012019 doi:10.1088/1757-899X/204/1/012019

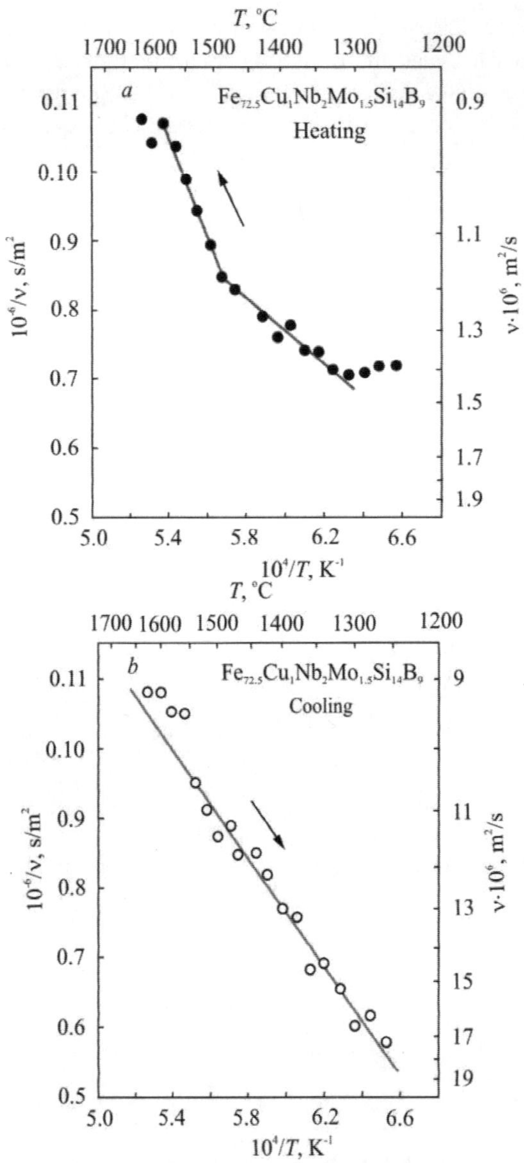

Figure 3. Dependences of the inverse kinematic viscosity of the melt 1/ν on the inverse absolute temperature 1/*T* in the course of heating (a) and cooling (b).

The results obtained indicate that for the multicomponent $Fe_{72.5}Cu_1Nb_2Mo_{1.5}Si_{14}B_9$ melt there is the critical temperature T_k = 1490 °C above which the melt can be converted into another structural state with a different magnitude of the activation energy of the viscous flow E_a. The cooling of the melt from the new structural state proceeds irreversibly. To compare two states of the melt, above and below the critical temperature, a ribbon 25 μm thick was cast from the melt prepared by two modes prior the following ultrarapid cooling. In the first mode, the melt of $Fe_{72.5}Cu_1Nb_2Mo_{1.5}Si_{14}B_9$ was heated to 1480 °C prior to casting a ribbon. In the second mode, the melt of the same ingot was heated to 1520 °C, then cooled to 1480 °C, and cast. At each temperature specified the exposure was 5 min.

Figure 4 shows the in-time changes in the temperature and initial permeability in the course of heat treatment at 550 °C for the cores fabricated from the ribbons that were subjected to two modes of preparation of the $Fe_{72.5}Cu_1Nb_2Mo_{1.5}Si_{14}B_9$ melt.

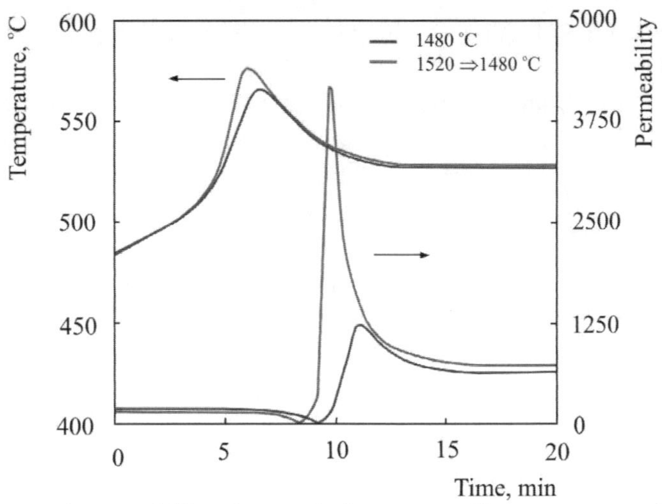

Figure 4. Temperature and permeability changes in the course of heat treatment of cores made from the ribbons undergone preparation of the $Fe_{72.5}Cu_1Nb_2Mo_{1.5}Si_{14}B_9$ melt by two modes.

The cores fabricated from the ribbons obtained by two modes of the melt preparation were subjected together to the heat treatment at the temperature 550 °C for 1 h after the crystallization peak. In Table 1, data are given on the permeability $\mu_{0.08}$ at the magnetic field strength of 0.08 A/m, maximal permeability μ_{max}, coercivity H_c of the magnetic hysteresis loop at the maximal magnetic field strength of 800 A/m. It is seen that the mode with overheating of the melt above the critical temperature provides higher magnetic properties.

Table 1 Magnetic properties of cores made from the $Fe_{72.5}Cu_1Nb_2Mo_{1.5}Si_{14}B_9$ alloy

Regime of melt preparation	$\mu_{0.08}$	μ_{max}	H_c, A/m
1480 °C	125000	882000	0.45
1520 → 1480 °C	164000	1024000	0.40

Summary

In the work, the temperature dependences of the kinematic viscosity of the multicomponent $Fe_{72.5}Cu_1Nb_2Mo_{1.5}Si_{14}B_9$ melt have been studied. A critical temperature is established above which the energy of activation of the viscous flow of the melt changes. In the course of cooling of the overheated melt, the temperature curves of the kinematic viscosity become linear within the given coordinates. In the amorphous ribbon obtained in the mode of preparation of the melt with overheating above the critical temperature, an enhanced value of the enthalpy of crystallization is found. After the heat treatments of the cores, the mode with overheating the melt above the critical temperature results in higher magnetic properties.

Acknowledgements

This work was supported by the Scientific researches of higher education institutions within the State task of the Russian Federation No. 2014/236.

References

[1] B.A. Baum, *Metal liquids*, Nauka, Moscow (1979).

[2] G. Herzer, "Nanocrystalline soft magnetic alloys", in *Handbook of Magnetic Materials*, **10**, edited by K.H.J. Buschow, Elsevier Science B.V. (1997) pp. 415–462.

[3] G. Kumar, T. Ohkubo and K. Hono, *J. Mater. Res.* **24** (2009) pp 2353–2360.

[4] U. Dahlborg, M. Calvo-Dahlborg, P.S. Popel and V.E. Sidorov, *Eur. Phys. J.* **B14** (2000) pp 639–648.

[5] A.L. Bel'tyukov, A.I. Shishmarin and V.I. Lad'yanov, *High Temperature*, **53** (2015) pp 302–304.

[6] V. Tsepelev, V. Konashkov, Y. Starodubtsev, Y. Belozerov and D. Gaipisherov, *IEEE Trans. Magn.* **48** (2012) pp 1327–1330.

[7] Ye.G. Shvidkovskiy, Certain problems related to the viscosity of fused metals, National Aeronautics and Space Administration, Washington (1962).

[8] V.V. Konashkov, V.S. Tsepelev, V.V. V`yukhin, A.M. Povodator and A.I. Podol`skaya, A Computer-Aided Plant for Studying the Kinematic Viscosity of High-Temperature Metallic Melts, *Instruments and Experimental Techniques*, **54** (2011) pp 284–285.

[9] V.S. Tsepelev, V.V. Konashkov, V.V. Vyukhin and A.M. Povodator, Russian Patent 2,366,925 (2009).

IOP Conf. Series: Materials Science and Engineering **204** (2017) 012020 doi:10.1088/1757-899X/204/1/012020

The effect of screw rotating speed on mass flow rate, temperature, viscosity, mooney scorch time anddie swell of cold feed rubber blending prepared by qsm 200 extruder machine

P Elmiawan[1], H Saryanto[1] and D Sebayang[2]

[1]Gajah Tunggal Polytechnic, Kel. Pasir Jaya, JatiUwung, Tangerang, Banten, Indonesia

[2]MercuBuana University, Jln. Meruya Selatan, Kembangan, Jakarta Barat, Indonesia

E-mail: elmiawan@gmail.com, hsaryanto@gmail.com, d_sebayang@hotmail.com

Abstract.The effect of screw rotating speed on cold feed rubber blending production were evaluated. The specimens were prepared by using QSM 200 extruding machine with variable screw rotating speed 4, 6, 8, 10, 12, 16 and 20 RPM. The temperature set in the screw and head of the barrel was around 80 °C by using Temperature Controll Unit (TCU), and was set around 70°C in zone 1 and 2. Mooney Viscometer was used to evaluate the viscosity and Mooney Scorch Time of cold rubber blending before and after extruding process. In addition, the dynamic rubber process analyzer was used to evaluate the Die Swell of rubber blending after extruding process. The result indicated that the increase of screw rotating speed has a significant effect to increase the flow rate, temperature and the viscosity of the rubber blending. Otherwise, the Mooney Scorch Time (MST) increases due to the decrease of the screw rotating speed. It does not have a significant effect to the die swell of rubber blending.

1. Introduction

Consumption of polymeric materials has greatly increased Over the past few decades due to their use in diverse industrial sectors [1]. The manufacturing process in this polymeric material undergoes several processes. One of the process is extrusion using extruder machine. It is one important process in polymeric material making. The extrusion process is used to create a semi finished product whose output is rubber blending. [2-6]

Screw rotation speed and temperature setting is needed in order to maintain the quality and quantty of the product. Controlling setting barrel temperatures and screw rotation speed directly influences process stability and product quality [7]. The various barrel temperature is the main challenge in the extrusion process [8] for the temperature dstribution in this process influence viscous and swelling of the extrusion output [9]. The extrusion result, however, define the characteristics of the rubber blending. The viscous characteristics is used to define the way it is processed on the next process. Time characteristic for schorch is used to define the time before the rubber blending is schorched.viscosity characterstics and time for schorch define how good is the material resulted from the extrusion process [10].

Content from this work may be used under the terms of the Creative Commons Attribution 3.0 licence. Any further distribution of this work must maintain attribution to the author(s) and the title of the work, journal citation and DOI.
Published under licence by IOP Publishing Ltd

IOP Conf. Series: Materials Science and Engineering **204** (2017) 012020 doi:10.1088/1757-899X/204/1/012020

In addition, it is also importat to know the score of the die swell parameter for in an extrusion process, rubber blending will experience changes in its viscosity, mooney scorch time (MST) and die swell [11]. Die swell is an important parameter to characterize the elasticity of the polymer in the extrusion process when it is melting [12]. Die swell measurement is the best qualitatve measurement of the flow behavior [13]. Cold feed extruder machine has better control of stock temperature than that of the hot feed one [4]. Therefore, it is important to conduct this study. This study is done to figure out the influence of screw speed variation and fixed temperature control toward the mass flow rate, viscosity, mooney schorch time and die swell using the cold feed extruder machine.

2. Experiment

The material, machine, measurement test equipment, experiment methods and analytical methods used in this study will be explained as follow:

2.1. Material.

This study used the cold material known as rubber blending prossesed by using a openmill machine whose output is sheet of rubber blending. This material composition Per Hundred of Rubber was a mixed of these following (table 1).

Table 1. Material Composition

MATERIAL	CATEGORY	COMPOSITION [PHR]
SIR-10	Natural Rubber	100
N-330	Carbon Black	30
HISIL 225	Silica	12
X-50 S	Coupling Agent	2
ZINC OXIDE	Activator	3
RHENOGRAN SBR-80	Bounding Agent	1.38

2.2. Machine

The extrusion process used QSM 200 machine whose diameter of the screw is 200 mm, screw length is 3200 mm, pin rows number is 12 (with 8 pins each), the speed rotation screw is 1.6-32 RPM, the Max output is 2000 kg/h, and Max temperatur is 125 °C. This is a cold feed extruder rubber blending machine.

2.3. Measurement Test Equipment.

Research support tools used are thermocouples, and stopwatch and weigher. Ektron EKT-2003M was the Mooney viscometer used to know the value of viscosity and Mooney Scorch Time. To analyze Die Swell, Mount Tech D-RPA 3000 was used.

2.4. Experimental Methods.

Cold Rubber blending is the compound processed by openmill machine which passes through the hopper to get to the next process – extrusion. Extrusion process was done in screw rotation variation of 4,6,8,10, 12 and 20 RPM. While the barrel temperature was divided to four zones, namely: screw, zone 1, zone 2 and head, with the temperature variance of 80 °C, 70 °C, 70 °C and 80 °C which is controlled by TCU

2.5. Analytical Methods.

Mass flow rate was analyzed after the extruding process. Viscosity and Mooney Schort Time (MST) was evaluated, done before and after the extruding process by using Mooney Viscometer Ektron EKT-

2003M and refering to standart ASTM D1646. Whereas die swell was evaluated after the extruding process by using D-RPA 3000 Mount Tech and refering to ASTM D6601

3. Result and Discusion

3.1. Analysis Of Screw Rotation Speed Toward Mass Flowrate.

The evaluation result of the effect of Screw Rotation Speed to the mass flow rate using QMS 200 machine was done by changing the screw speed is shown in figure 1

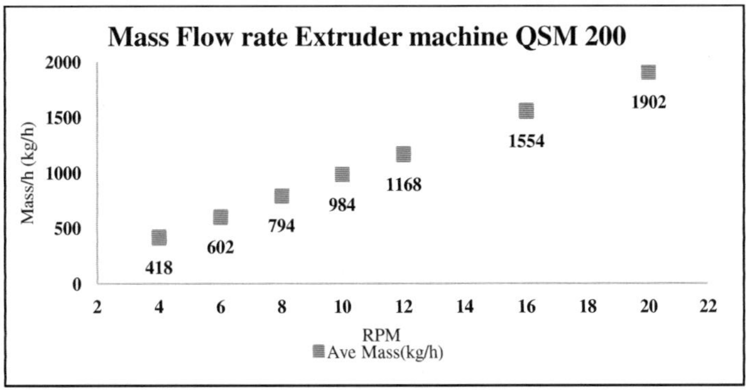

Figure 1. Graph result screw rotating speed on mass flow rate

Figure 1 shows the minimal score of 418 kg/h and the maximum score of 1902 kg/h. in terms of screw speed and mass flow rate, it showed that the increase of the screw rotating speed, will increase the mass flow rate.

3.2. Analysis Of Screw Rotating Speed Toward Temperature.

The evaluation of the effect of screw rotating speed toward the material temperature resulted by the QMS 200 machine done by changing the screw speed is shown by figure 2

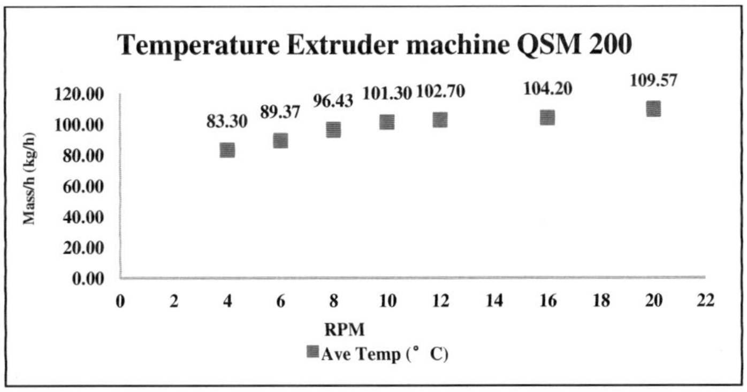

Figure 2. Graph result screw rotating speed on Temperature

Figure 2 showed the result of the evaluation which is done by increasing the screw speed. It showed 83.3 °C as the lowest temperature and 109.57 °C as the highest temperature. In this case, the increase of the screw rotating speed increases the temperature.

3.3. Analysis Of Screw Rotating Speed Toward Viscosity.

The viscosity test on the pre-extruding process material, which is done in 5 minutes - refering to ASTM D1646 - showed 55.62 mooney. After the extrussion process, the viscosity test was done. The test is aimed at figuring out the viscosity value. The test showed 54.02 mooney as the minimum value and 54.53 mooney as the maximum value. Viscosity test result on the extruded material can be seen in figure 3.

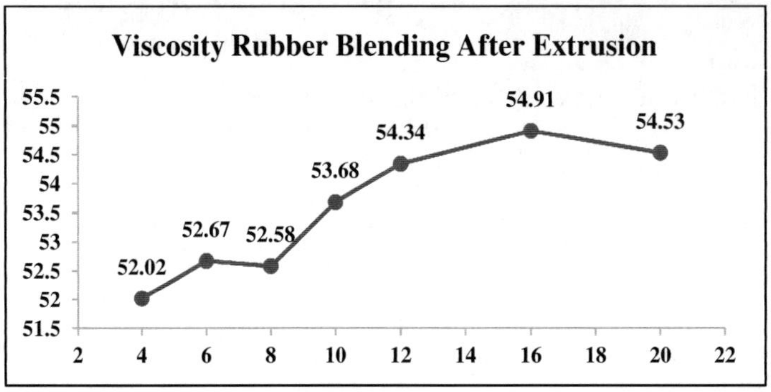

Figure 3. Graph result screw rotating speed on viscosity after extrusion

Based on the result shown in figure 3, screw rotating speed influenced the viscoosity. The increase of the screw rotating speed, will increase the viscosity.

3.4. Analysis Of Screw Rotating Speed Toward MST

Before the extrusion, the MST showed 9.12 minutes. While after the extrusion, the MST showed 9.28 minutes as the minimum scorch time and 10.32 minutes as the maximum. it is shown by figure 4

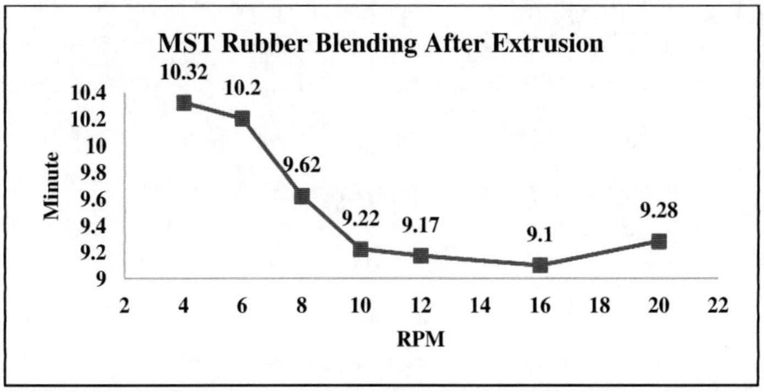

Figure 4. Graph result screw rotating speed on MST after extrusion

Based on the data shown in the figure above, screw rotating speed influenced the MST. The higher the RMP the lower the MST would be.

3.5. Analysis Of Screw Rotating Speed Toward Die Swell.

The die swell test result can be seen in figure 5. Die swell test showed the minimum score of 0.3379 and maximum score of 0.3449

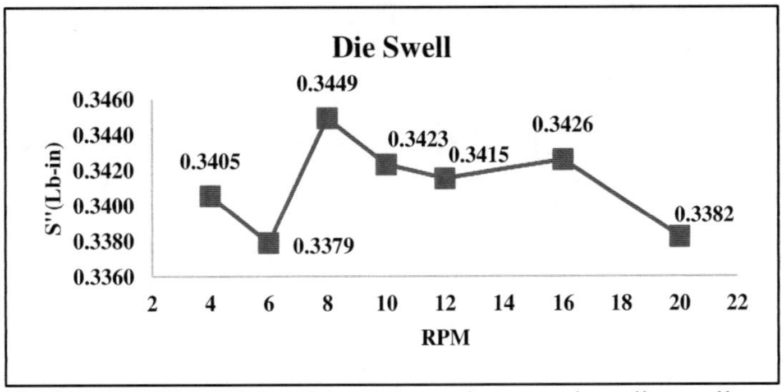

Figure 5. Graph result screw rotating speed on die swell

In this case, screw rotating speed did not influence the die swell score since die swell has the unstable score. The changed screw rotating speed score did not give significant effect toward die swell score.

4. Conclusion

The result of the experiment showed that the increase of the screw rotating speed increased the mass flow rate, temperature, and viscosity. In another way, the increase of the screw rotating speed decreased the mooney scorch time. However, the screw rotating speed did not influenced the die swell score at all. To conclude, based on that findings, the characteristics of the extruded rubber blending can be identified. Therefore, the exact screw rotating speed can be determined to meet the required extrusion result.

References

[1] Vera-Sorroche J, Kelly A L, Brown E C, Gough T, Abeykoon C, Coates P D and Price M 2014 The effect of melt viscosity on thermal efficiency for single screw extrusion of HDPE. *Chemical Engineering Research and Design.* **92.**11 pp 2404-2412.

[2] Stevens M J andCovas J 2012 *Extruder principles and operation*(Springer Science and Business Media)

[3] Rauwendaal C 2014 *Polymer extrusion*(Carl HanserVerlag GmbH Co KG)

[4] Kostic M M and Reifschneider L G 2006 Design of extrusion dies. *Encyclopedia of Chemical Processing* DOI, **10**, pp 633-649.

[5] Michaeli W andMichaeli W 2003 *Extrusion dies for plastics and rubber* (Carl HanserVerlag GmbH and Co. KG) pp. 13-17

[6] Testi S, Bottomley Aand Longoni A 2016 *Process and plant for producing tyres* (Washington DC: U.S. Patent and Trademark Office) U.S. Patent No. 9,365,006

[7] Vera-Sorroche J, Kelly A, Brown E, Coates P, Karnachi N, Harkin-Jones E andDeng J 2013 Thermal optimisation of polymer extrusion using in-process monitoring techniques.*Applied Thermal Engineering.* ***53*** 2 pp 405-413

[8] Godovsky Y K 1992*Thermomechanics of Glassy and Crystalline Polymers In Thermophysical Properties of Polymers*(Berlin :Springer)pp 127-162

[9] Michaeli W andMichaeli W 2003 *Extrusion dies for plastics and rubber* (Carl HanserVerlag GmbH and Co. KG) pp 287-304

[10] Dick J S 2001*Rubber Technolog* (Munich:Hamser)

[11] Mokhtar H, Nordin R, Shamsuddin S and Noriman N Z 2014 White fly ash filled natural rubber compounds: effect on mooney scorch time, swelling behavior, hardness and resilience. *American-Eurasian Journal of Sustainable Agriculture* pp 99-104.

[12] Liang J Z 2008 Effects of extrusion conditions on die-swell behavior of polypropylene/diatomite composite melts. *Polymer testing* **278** pp 936-940.

[13] Rides MandAllen CRG1996 Intercomparison of extrudate swell measurements using capillary extrusion rheometers for a polyethylene melt*NPL Report CMMT* A 9

The effects of screw rotation speed on viscosity, mooney scorch time and dieswell of hot-feed blending rubber-compound prefared by gs/w 250 extruder machine

A Zohari[1], H Saryanto[1], and D Sebayang[2]

[1]Polytechnic Gajah Tunggal, Kel. Pasir Jaya, Jati Uwung-Tangerang, Banten-Indonesia
[2]Mercu Buana University, Jln. Meruya Selatan-Kembangan-Jakarta Barat, Indonesia

E-mail: zohari2980@gmail.com, hsaryanto@gmail.com, d_sebayang@hotmail.com

Abstract. This paper is an experimental study that aimed at figuring out the effect of screw rotation speed on the viscosity, mooney scorch time and dieswell of rubber-compound blending prepared by GS/w 250 Extruder Machine to determine the appropriate parameter setting for the expected rubber-compond specification. The screw rotation speed was set at 10, 20, 40 RPM and the three-zone temperature of the screw extruder was set by the TCU. Viscosity and Mooney Scorch Time (MST)of rubber-compound were evaluated before and after extruding processusing the Mooney Viscometer test equipment. The Die Swell of the rubber-compound were evaluated after blending process using the dynamic rubber process analyzer. The result showed that the Viscosity and Mooney Scorch Timeof rubber-compound before extuder process was around 56.59 mooney and 19.69 minute. After extruding process, the value of Mooney Viscosity was consecutively around 51.55, 51.25, 50.55 mooney and the value of Die Swell was consecutively around 0.493, 0.536, 0.553Lb-in. The value of Mooney Scorch Time was consecutively around 19.00, 15.47, 17.37 minute. The results indicates that, the increase of screw rotation speed decrease the Viscosityand increase the Die Swell of the rubber-compound.

1. Introduction

The consumption of polymer have been widely used in rubber industry. Polymer has mechanical and thermal properties of interest. Several polymer have ductile and resistant to acid, but will be swelling when in contact with gasoline, oil or grease. Therefore, for special application such as tyre its toughness is not sufficient. The strength of a polymer can be considerably improved by blending in a rubber [1-8]. This blends can be made in a reactor process on an extruder [9].This blends produce very ductile materials, commonly called rubber-compound. The extruder is arguably the single most important piece of polymer processing machinery [10,11]. The extrusion process are controlled by setting barrel temperatures and screw rotation speed [12].

The effects of screw rotation speed on extruding process has been studied extensively. One study focussed on HDPE rheology in a given condition whose result reflected the dependency of the variation in radial melt temperatures on screw rotation speed,and polymer viscosity [13]. Other studied the effectsof screw speed on the quality of devulcanized rubber whose result showed that the screw speed affects the percent of devulcanization, especially at low and middle levels of barrel temperature. So, with the increase of screw speed devulcanization process becomes more efficient [14].

There are three correlated method in evaluating rubber-compound. Mooney Viscosity, Mooney Scorch time, and die swell. Mooney Viscosity test is the most popular test method to evaluate the character

Content from this work may be used under the terms of the Creative Commons Attribution 3.0 licence. Any further distribution of this work must maintain attribution to the author(s) and the title of the work, journal citation and DOI.
Published under licence by IOP Publishing Ltd

polymers and rubber materials.Rubber-compound viscosity plays vital role in deciding its processing behaviour,whereas the deviation in viscosity will critically alter its processibility such as in calendering, extruding or injection moulding. So, it is necessary that viscosity parameter be maintained within specified limits. The second one, Mooney Scorch, is one of the most useful tests to determine starting of cure of rubber-compounds known as scorching behavior, providing essential data for designing and controlling production processes as well as checking material consistency.It indicates how long a rubber-compound can remain warm for processing before it loses its ability to change its shape. Often once a rubber-compound is scorched, it can no longer be processed. Therefore, it is important to measure the scorch time and adjust it based on the processing conditions.In a quality control environment, evaluating the scorch time ensures that no time is wasted processing or molding material that will not create a good product for premature scorch can lead to underfilled mold cavities as the material loses its ability to flow before completely filling the space.The last one is die swell test. Die swell occurs when an extrudate has larger diameter than die [10]. It is an important parameter for characterization of elastic properties during extrusion of polymer melts, and the degree of swell degree is usually expressed by die-swell ratio [15]. The measurement of die swell is best as a qualitative measure of the flow behavior [16].
Several rubber-compound processing will be change its characteristic. Therefore, the tests mentioned before has to be done to evaluate of rubber-compound characteristic in order to ensure consistency quality of the finished goods. Based on that reason, this paper studied the effects of screw rotation speed on viscousity, mooney scorch time and dieswell of rubber-compound using hot-feed Extruder Machine.

2. Materials, Extruder Machine, Measurement Test Equipment and Experimental Processes

Material, apparatus and experimental processes will be explained below:

2.1. Materials

The materials used in this experiment was hot-feed rubber-compound whose composition is shown in Table 1.

Table 1. Rubber-compound materials

Material	Kategori	Formulation (phr)
MR-01	Natural Rubber	75
BR-1220	Polimer Rubber	15
SBR-1502	Polimer Rubber	10
HISIL 225	Silica	15
N-234	Carbon Black	40
X-50 S	Coupling Agent	3

2.2. Extruder Machine

A co-rotating single-screw extruder (Model No. GS/w 250, TROESTER GmbH & Co. KG., Am Brabrinke 1-4 Hannover, GERMANY) was used in this study. The extruder‘s screw diameter is 250 mm with an L/D ratio of 8.

2.3. Measurement Test Equipment

Refering to ASTM D1646, a mooney Viskometer (Model Ektron EKT-2003M) was used to measure the viscosity and mooney scorch time test before and after extruding process. In evaluating the die swell, based on ASTM D6601, the Dynamic Rubber Process Analyzer (Model Mon Tech D-RPA 3000) was used.

2.4. Experimental Process

The preheating extruding process was used in order to produce hot-feed rubber-compound which was then used to feed the experimented extruding machine. Hot-feed compound then entered the barrel whose temperature was set in three zones: feed (80°C), compression (70°C) and metering (80°C). The screw rotation speed variations were: 10, 20 and 40 rpm delivered the compound to the head extruder. The viscosity, mooney scorch and dieswell testing were done to test the rubber-compound.

3. Result and Discussion

The experimental result is as follow:

3.1. Mooney Viscosity

The Viscosity of hot-feed rubber-compound before extruding process was around 56.59 mooney. The viscosity of rubber-compound was affected by screw rotating speed as is shown by figure 1.

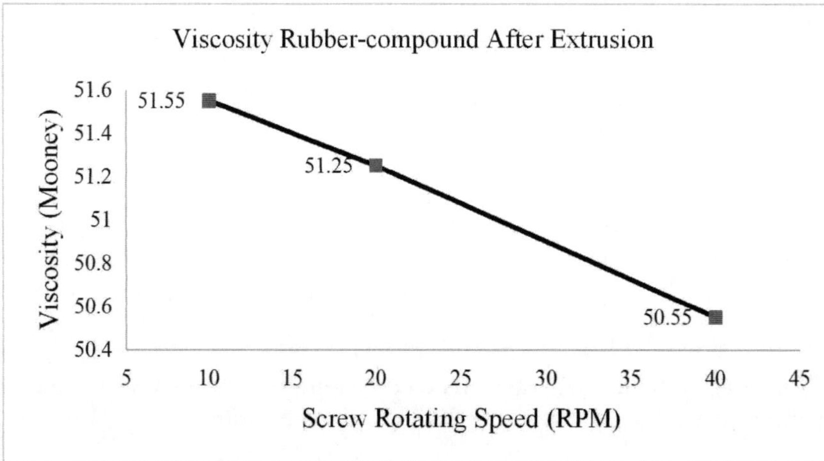

Figure 1. The effect screw rotating speed on viscosityof rubber-compound

Figure 1above shows that the minimum screw rotating speed was at 10 rpm, the viscosity was 51.55 mooney. When the maximum speed was applied (at 40 rpm), the viscousity changed to 50.55 mooney. It shows that the increase of screw rotation speed will decrease the viscosity of rubber-compound.

3.2. Mooney Scorch Time

The Mooney scorch time of hot-feed rubber-compound before extruding process was around 19.69 minute. The mooney scorch time after extruding process with screw rotating speed variation is shown in figure 2.

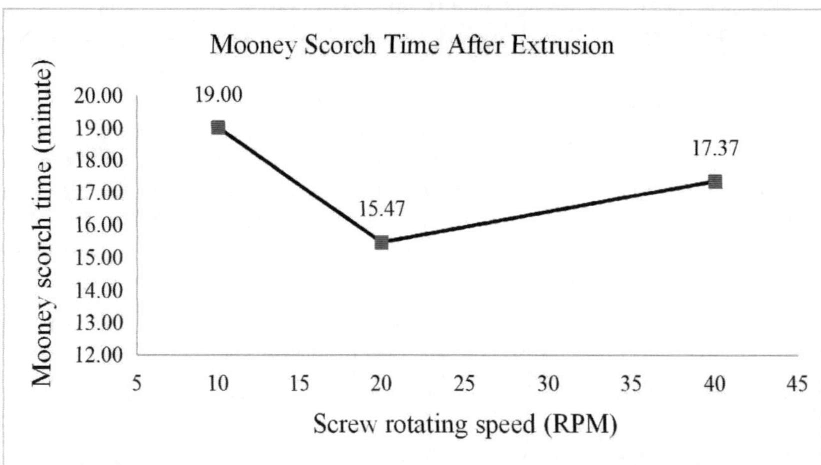

Figure 2. The effect screw rotating speed on mooney scorch time of rubber-compound

Figure. 2above shows that the minimum value of mooney scorch time is 15.47 minute when the screw rotating speed was at 20 rpm, and the maximum value of mooney scorch time is 19.00 minute when the screw rotating speed was at 10 rpm. It shows that the mooney scorch time of rubber-compound does not affected by the increase of screw rotation speed.

3.3. Die Swell.

The die swell value after extruding process with screw rotation speed variation is shown in figure 3.

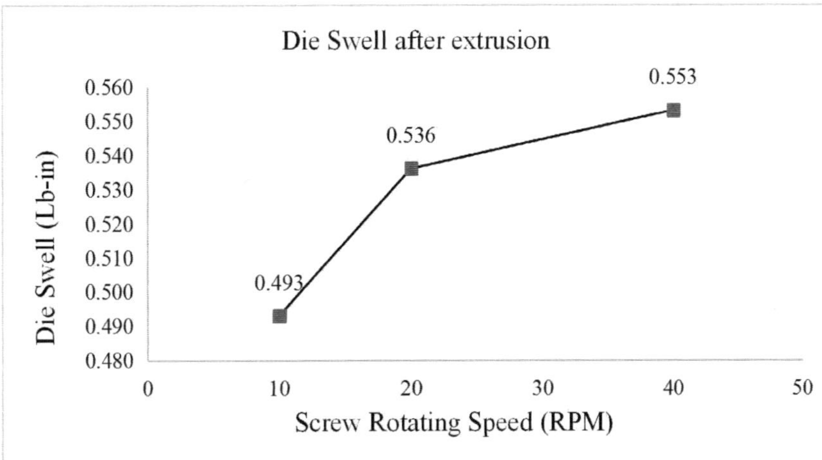

Figure 3. The effect screw rotating speed on die swell of rubber-compound

Figure. 3 above shows that the value of die swell was consecutively around 0.493, 0.536 dan 0.553 Lb-in. It shows that increasing screw rotation speed increase the value of die swell in linear approach.

4. Conclusion

Extrusion process affected the rubber-compound characteristics. It can be seen through the changes of viscosity and die swell value due to the changes of the screw rotation speed where the increase of screw rotation speed does not only change the viscosity value, but also increase the die swell value of rubber-compound.The above changes can be used to define the appropriate extrusion's process parameter in order to meet the required specification of the rubber-compound output.For the following experiments, it is suggested to add the variation rotation speed of screw.

References

[1] Taşdemir M, 2004*Journal of applied polymer science*, **93(6)**, 2521-2527
[2] Kim S D, Choi Y, Choi W, Choi C, and Chun Y S, 2012 Effect of Ethylene - Propylene Copolymer Composition on Morphology and Surface Properties of Impact propyleneCopolymer *Macromolecular Symposia* **312** No 1 pp 27-33
[3] Mehta, Sameer, Mirabella, Francis M, Rufener, Karl, and Bafna 2004 Thermoplastic olefin/clay nanocomposites: morphology and mechanical properties *Journal of Applied Polymer Science,* **92(2)**, pp 928-936
[4] Zhao, Chengzhe, Wu, Guangfeng, Zhou, Chao, Yang, Haidong, Zhang and Huixuan 2006 Independence of the brittle–ductile transition from the rubber particle size for impact - modified poly (vinyl chloride), *Journal of Polymer Science Part B: Polymer Physics*, **44(4)**, pp 696-702
[5] Zhang, Ling, Chunzhong L, and Rui H, 2004 Toughness mechanism in polypropylene composites: polypropylene toughened with elastomer and calcium carbonate *Journal of Polymer Science Part B: Polymer Physics* **42.9** pp 1656-1662
[6] Yu, Long, Dean, Katherine, Li, and Lin 2006 Polymer blends and composites from renewable resources, *Progress in polymer science*, **31(6)**, pp 576-602

[7] Yu, Zhong-Zhen, Yang, Ming-Shu, Dai, Shao-Cong, Mai and Yiu-Wing 2004 Toughening of recycled poly (ethylene terephthalate) with a maleic anhydride grafted SEBS triblock copolymer. *Journal of applied polymer science*, **93(3)**, pp 1462-1472

[8] Jiang, Long, Wolcott, Michael P, Zhang and Jinwen, 2006 Study of biodegradable polylactide/poly (butylene adipate-co-terephthalate) blends. *Biomacromolecules*, *7*(**1**), pp 199-207

[9] Tadmor, Zehev, Gogos and Costas G, 2013 *Principles of polymer processing*, John Wiley & Sons44.

[10] Stevens M J and Covas J 2012 *Extruder principles and operation* (Springer Science & Business Media)

[11] Rauwendaal C 2014 *Polymer extrusion* (Carl Hanser Verlag GmbH Co KG)

[12] Vera-Sorroche J,Kelly A, Brown E, Coates P, Karnachi N, Harkin-Jone E, and Deng J2013 Thermal optimisation of polymer extrusion using in-process monitoring techniques *Applied Thermal Engineering* **53.2** pp 405-413.

[13] Vera-Sorroche J, Kelly A L, Brown E C, Gough T, Abeykoon C, Coates P D, and Price M2014 The effect of melt viscosity on thermal efficiency for single screw extrusion of HDPE *Chemical Engineering Research and Design* **92.11** pp 2404-2412

[14] Yazdani, Hamid, Karrabi, Mohammad, Ghasmi, Ismaeil, Azizi, Hamed, Bakhshandeh and Gholam R, 2011 Devulcanization of waste tires using a twin‐screw extruder : The effects of processing conditions, *Journal of Vinyl and Additive Technology*, *17*(**1**), pp 64-69

[15] Liang J Z 2008 Effects of extrusion conditions on die-swell behavior of polypropylene/diatomite composite melts*Polymer testing* **27.8** pp 936-940.

[16] Rides M and Allen C R G1996 Intercomparison of extrudate swell measurements using capillary extrusion rheometers for a polyethylene melt*NPL Report CMMT A9*

The use of natural sand from lampusatu beach, kabupatenmerauke, papua for mixed asphalt concrete

D D M Huwae[1], L. R. Parera[2], Alpius[3] and J Tanijaya[3]

[1] Politeknik Negeri Ambon, Ambon, Indonesia
[2] University of Pattimura, Ambon, Indonesia
[3] Paulus Christian University of Indonesia, Makassar, Indonesia

E-mail:david_daniel_amb63@yahoo.com, lykiera_amb64@yahoo.com, alpius_andre @live.com, jonie.tanijaya@gmail.com

Abstract. The natural sand from LampuSatu Beach, KabupatenMerauke, Papua, is often used as building material by local people. This research aims to test the use of this natural sand for mixed asphalt concrete. Asphalt Concrete Wearing Course (AC-WC) with bitumen penetration 60/70 and variations of asphalt content 5%, 6%, 7%, 8% and 9% are used in this research. Testing result shows the stability 1372.17 kg, VIM 3.95%, VMA 16.81%, Flow 4.14 mm, and MQ 332.89 kg/mm. It has fulfilled the standard requirements set by Indonesian Director General for Highways. The percentage of Optimum Asphalt Content (OAC) is 7% and the Index of Retained Strength (IRS) is 107.84% (≥ 75%). These indicate that the mixture can fulfill the Stability and Marshall Immersion Test.

1. Introduction

There are three types of asphalt concrete based on its function, namely asphalt concrete as wear layers, binder layer and base course. The material of mixed asphalt concrete consists of aggregate, filler and bitumen. Filler usually consists of fly-ash serving as a binding gap for both material and air in the mix. This can increase the mass density of the mixture. The total amount for the use of filler in asphalt mixture is at least 1% of the total weight aggregate. The gradation of combined aggregate for mixed asphalt concrete layer must fulfill the limits and should be outside the "restriction zone" in accordance with the provisions of Indonesian Director General for Highways. In Indonesia it is typically to use hard to moderate asphalt penetration 60/70 and 80/100. The performance of asphalt concrete can be determined through Marshall Test (Marshall Characteristics). Marshall Test consists of some parameters, such as stability and melting (flow), Void In Mix (VIM), the levels of the cavity filled with asphalt (Voids Filled Asphalt, VFA), cavities in the aggregate (Voids In Mineral Aggregate, VMA), Marshall Quotient (MQ), and Compacted Aggregate Density (CAD). Marshall Test activities include: a) preparation of the specimen, b) determining the mass density of the specimen, c) checking the value of stability and flow, and d) the calculation of volumetric properties of the mixture.

The development of road construction at KabupatenMerauke, Papua, as one of the provinces in Indonesia, requires a very enormous expense because the pavement material should be brought from outside. There is not enough adequate resources at KabupatenMerauke. The natural sand from LampuSatu Beach, KabupatenMerauke, Papua, is often used as the building material by local people. However it has the hidden potential to be used as fine aggregate for mixed asphalt concrete. The natural sand materials from various places have been commonly tested and used as fine aggregate for

Content from this work may be used under the terms of the Creative Commons Attribution 3.0 licence. Any further distribution of this work must maintain attribution to the author(s) and the title of the work, journal citation and DOI.
Published under licence by IOP Publishing Ltd

mixed asphalt concrete. The use of natural sand from the local place can reduce the expenses needed for road construction.

Previous researches show that the natural sand from Central Kalimantan [1], natural sand from Gorontalo [2], natural sand from Podi River, Palu [3], natural sand from Kampar [4], natural sand from Bakau Beach, Seruyan city, Central Kalimantan [5] and natural sand from Kampar River, West Sumatra (Alfian, 2015) have fulfilled the requirements to be used as fine aggregate in mixed asphalt concrete, specifically for AC-WC. Therefore this research is conducted to test on the possibility of using natural sand from Lampu Satu Beach, Kabupaten Merauke, Papua. Asphalt Concrete Wearing Course (AC-WC) with bitumen penetration 60/70 is used in this research. The variations of bitumen content used are 5%, 6%, 7%, 8% and 9%. The characteristic of the natural sand from Lampu Satu Beach as fine aggregate is tested based on Marshall Test. The results of this research is expected to be the source of information for stakeholders of road construction in Papua, especially in terms of using natural material as an aggregate for mixed asphalt concrete, as well as being a reference for designing the mixture of AC-WC by using the natural sand from Lampu Satu Beach as fine aggregate.

2. Literature review

The minimum thickness of base layer (AC-Base) is 7.5 cm with a maximum aggregate size of 37.5 mm [6]. The hot mix asphalt concrete must fulfill some certain conditions to meet the criteria for specific traffic conditions [7]. It must be able to withstand long-term deformation and cracking due to fatigue loading, easy to implement, satisfactory compacted, water-proof, abrasion resistant due to the effect of traffic, air and water, contributed to the strength of pavement structure, easy to maintain, and most importantly, economical.

The performance of mixed asphalt concrete is dependent to the characteristics of the used aggregate, such as porosity, gradation and size distribution, absorption, shape and surface texture, rupture strength, modulus of elasticity, and the presence of substances that can damage the mixture [8]. Aggregate can be classified based on its volume density, origin size and mass density [9]. It is classified as sand-gravel (1520-1680 kg/m^3), light-weight aggregates (<1.120 kg/m^3) and heavy-weight aggregate (> 2.080 kg/m^3) based on its mass density. The aggregates are classified into: coarse aggregate (retained sieve number 4), fine aggregate (through sieve number 4) and filler ($\geq$ 75% by weight through sieve number 200) based on its origin size. Fraction of coarse aggregate for asphalt mixture has a grain retained on the sieve number 8 (2.36 mm) through wet sieving, clean, hard, durable, and free from clay. The selection of the type of asphalt depends on the type of construction and the climate of a region. Asphalt texture will be hard and brittle at low temperatures and become soft at high temperatures [10]. The amount of bitumen content in pavement mixture ranges from 4-10% by weight of the mixture, or 10-15% of the volume of the mixture [11]. The asphalt classification is based on the value of penetration bitumen at a temperature of 25^{0}C, which is ranged from 4^0 to 30^0.

3. Methodology

Coarse aggregates were taken from Bili-Bili, South Sulawesi. Fine aggregates were taken from LampuSatu Beach, KabupatenMerauke, Papua. Asphalt binder was obtained from the Laboratory for Testing Road and Bridge Material at Indonesian Director General for Highways, South Sulawesi. Tonasa Cement was used as filler and taken from South Sulawesi.

The testing of AC-WC includes Void in Mix (VIM), stability, flow, Marshall Quotient (MQ) and Void in Mineral Aggregate (VMA). The Marshall Test was initially performed to predict the optimum asphalt content. Then, the specimen was prepared based on the predicted optimum asphalt content. Three specimens with the asphalt content higher than the optimum value and two specimens with asphalt content lower than the optimum value were also made with having each different value 0.5%. Three specimens were made for each asphalt content. The standard size for specimen used in testing was 63.5 mm (2.5 in). The adjustment for each the mixing temperature and compaction process were performed when the asphalt viscosity value reached about 170 ± 20 centistokes and 280 ± 30 centistokes, respectively.Compaction process was performed by applying 2 x 75 blows using Marshall compacting hammer. Then the specimen was kept in room temperature for 24 hours. Finally, the specimen was measured and weighed in air and water. The specimen was also measured in saturated surface dry condition. Then, they were allowed to cure at temperature 600C for 30 minutes and ready to test. The testing parameters were stability value, flow value, MQ, VIM and VMA. The optimum asphalt content (i.e. the asphalt content that reaches all design criteria for AC-WC) could be determined. In this research, the optimum asphalt content was then called the Optimum Asphalt Content (OAC).

4. Results and discussion

4.1 Material characteristic and Marshall test result

The characteristic of fine and coarse aggregates were tested. These included the gradation and inspection of physical properties. The testing result is shown in table 1. Testing results indicate that both fine and coarse aggregate used for mixed asphalt concrete AC-WC has fulfilled and met the specification requirements of Indonesian Director General of Highways (2010).The asphalt content used in AC-WC mixture was the Optimum Asphalt Content (OAC) which has reached all design criteria according to the standard Marshall parameter 7.0% (figure 1). Marshall Test result that illustrates the value of VIM, stability, flow, MQ and VMA is shown in Table 1. The graph results are illustrated in figure 2. The testing result shows that the VIM values had been decreasing along with the increasing of asphalt content which is varied from 5.0% to 9.0%. The increasing of asphalt content could make the voids between the aggregate particles in the mix more and more filled with asphalt so that the value of VIM would increase.

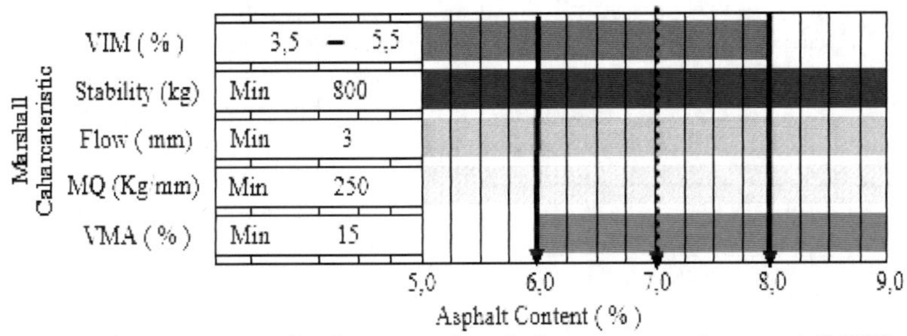

Figure 1. OAC analysis diagram of asphalt concrete mixtures AC-WC

Table 1. Marshall characteristic of asphaltconcrete

% asphalt	VIM	Stability	Flow	MQ	VMA
5,0	4,94	907,97	3,25	303,55	13,79
6,0	4,17	1239,32	3,68	349,27	15,07
7,0	3,95	1372,17	4,14	332,89	16,81
8,0	3,84	1659,74	4,26	397,29	18,61
9,0	3,34	1670,30	5,31	331,52	20,04

Stability test indicates that the stability value increases along with the asphalt addition, as more and more asphalt was used, the amount of asphalt that surrounds the aggregate would be growing. Therefore the asphalt covers that was created can be a good binder for aggregate to increase its stability. The flow value shows that the increasing of bitumen in the asphalt aggregate mixture can make the asphalt cover become thicker. Therefore it would increase the flow value as a function of the asphalt is for flexibility.The MQ results indicate that the asphalt content would decrease on 7.0%, increase on 8.0% and decrease back on 9.0%. However, the values of MQ entirely still fulfill the standard of Indonesian Director General of Highways.The standard value of MQ for heavy traffic should be at least 250 kg/mm. It can be deduced that the increasing of MQ affects both the increasing of stability and decreasing of the flow. Otherwise the decreasing of MQ cause the decreasing of stability and increasing of the flow. The VMA value increases with the increasing of the asphalt content in the mixture. The increasing of asphalt content whose function is to fill cavities in the aggregate can filled the void within asphalt. Therefore the VMA value would be increasing.

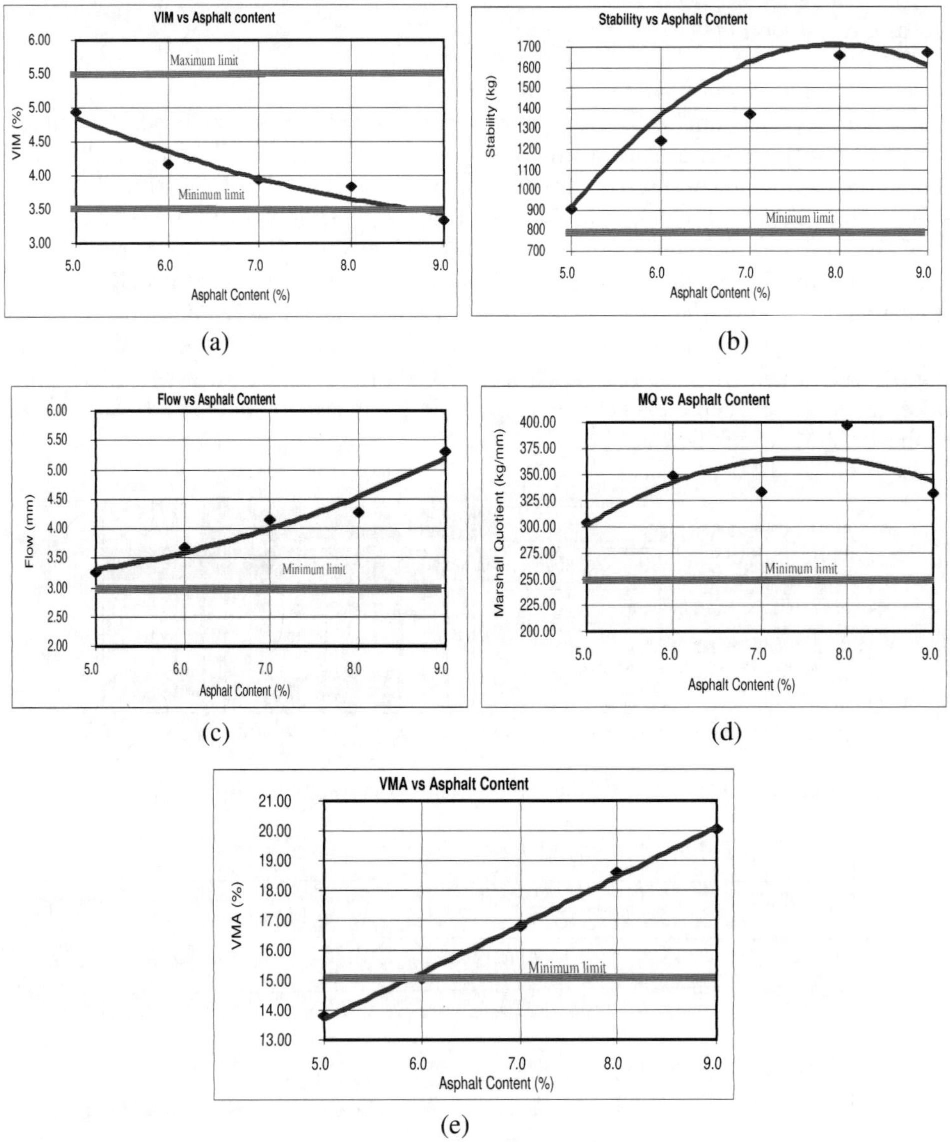

Figure 2. Marshall test result

4.2 Marshall immersion test result

The Index of Retained Strength (IRS) obtained from Marshall Immersion Test is 107.84%. This IRS value has fulfilled all the standard criteria (≥ 75%) specified by Indonesian Director General of Highways (2010). This indicates that the pavement using natural sand from LampuSatu Beach, Merauke, Papua with asphalt content of 7.0% has the resistance of temperature and water immersion, during the service time. This natural sand could not absorb the water in large amount. Hence even though the pavement is immersed in the water, the amount of water could not be absorbed directly into

IOP Conf. Series: Materials Science and Engineering **204** (2017) 012022 doi:10.1088/1757-899X/204/1/012022

the pavement. The resistance and durability of the pavement can be maintained if it has the proper, routine and periodic maintenance.

5. Conclusion

There are some conclusions that can be gathered from the data analysis. The testing results of natural sand from LampuSatu Beach, KabupatenMerauke, Papua, indicates that its characteristics has fulfilled the standard set by Indonesian Director General of Highways. Therefore it can be used as fine aggregate in road pavement.The percentage of Optimum Bitumen Content (OBC) which can be used to mix the asphalt concrete using the natural sand from Lampu Satu Beach as fine aggregate is varied from 6.0% to 9.0%. To optimize the mixing process, the midpoint values of OBC should be taken as 7%.The Index of Retained Strength (IRS) (107.84% ≥ 75%) indicates that the mixture can provide the stability and also fulfill the Marshall Immersion Test value. The mixture using natural sand from Lampu Satu Beach as fine aggregate has the resistance to the temperature changing, climate, or even when fully immersed in water.

References

[1] Nono. (2009). The Use of Natural Sand from Central Kalimantan for Asphalt Concrete Surface Layer. *Puslitbang for Road and Bridge.* **26**: No. 1

[2] Achmad, Fadly. (2010). Overview Aggregate Characteristic For Hot Asphalt Mixture (A Case Study of Some Quarry In Gorontalo). *Journal of Saintek.* **5**: No.1

[3] Setiawan, Arief. (2011). The Use of Natural Sand from Podi River As Fine Aggregate in Hot Rolled Sheet-Wearing Course (HRS-WC). *Journal of SMARTek.* **9**: No. 2 pp 109–121

[4] Santosa, L., and Alwinda, Y. (2013). The Use of Natural Sand for Asphalt Mixture Type AC-WC with Marshall Test Based on Specification from Indonesian Director General of Highways 2010. Research Report of Civil Engineering Department. Riau University, Pekanbaru

[5] Bestari, Akhmad. (2013). The Use of Natural Sand from Bakau Beach For Mixed Asphalt Concrete Type Hot Rolled Sheet (HRS). *Journal of Anterior.* **12**: No. 2 pp 13 – 22

[6] Anonym. (2010). *General specifications for road and bridge highways.* Indonesian Director General of Highways, Jakarta

[7] Read, J., and Whiteoak, D. (2003). *The bitumen Hand-book 5th edition.* London: Thomas Telford Publishing

[8] Siegfried, Yamin, HRA. and Fransisko, S. (2014). Optimizing Utilization of Local Aggregates from Kabupaten Talaud for Pavement Materials. Kolokium Road and Bridge 7th. ISBN: 978-602-264-038-7, Pusjatan, Bandung

[9] Susilorini, R.S., and Sambowo, K.A. (2011). *Advanced concrete technology and concrete Durability 2nd Edition.* Semarang: Surya Perdana Semesta (SPS) Publishing

[10] Yeni, S. M, dan Widayat, D. (2014). Determination of Bitumen Performance Grade. Kolokium Road and Bridge 7th. ISBN: 978-602-264-038-7, Pusjatan, Bandung

[11] Sukirman, S. (2003). *Hot Mix Asphalt Concrete.* Jakarta: Granit Publishing

IOP Conf. Series: Materials Science and Engineering **204** (2017) 012023 doi:10.1088/1757-899X/204/1/012023

Phase transformation of NiTi alloys during vacuum sintering

Wang Jun[1] and Kuang Hu[2]

[1]School of Metallurgy and Material Engineering, Chongqing University of Science and Technology, University City,Shapingba District, Chongqing, China

[2]School of Material Science and Engineering, Chongqing University, Chongqing, China

E-mail:wangjunpaper@126.com,574138516@qq.com

Abstract: The aim of this study is to ascertain the Phase transformation of NiTi alloys during vacuum sintering. NiTi shape memory alloys (SMA) of atomic ratio 1:1 were prepared through press forming and vacuum sintering with the mixture of Ni and Ti powders. Different samples were prepared by changing the sintering time and the sintering temperature. Phase and porosity of the samples were investigated by X-ray diffraction (XRD) and scanning electron microscope (SEM). The results show that in the process of sintering $NiTi_2$ and Ni_3Ti phases are formed firstly and then transform into NiTi phase. The quantity of $NiTi_2$ and Ni_3Ti phases gradually decreased but not eliminate completely with increase of sintering time. The porosity of specimen sintering at 900℃ decreases slightly with increase of sintering time. With increase of sintering time the porosity of specimen sintering at 1050℃ decreased firstly and then increased because of generation rich titanium liquid in the process of sintering.

1. Introduction

NiTi shape memory alloy of atomic ratio 1:1 has the advantages of good shape memory effect, superelasticity and biocompatibility, etc.[1]. These outstanding properties allow commercial applications of NiTi SMAs in the fields of biomedicine[1-3], intelligent machine[4.5] and aerospace [6].

The advantages of preparing NiTi alloy by powder metallurgy technology are[7]: (1) the compositions of sintered body are homogenous and esy to control; (2) it is easy to add alloy elements and prepare composite material; (3) it is easy to fulfill near-net-shaped forming of complex components, to reduce the amount of post machining. Thus, powder metallurgy technology has become an important research direction of preparation method of NiTi alloy. Presently, researches in this area mostly focus on preparation of porous NiTi shape memory alloy[8-11], while there are few research reports on preparation of dense NiTi alloy by powder metallurgy technology. There are two problems need to be solved. The first is existence of $NiTi_2$ and Ni_3Ti impurity phases in the sintered compact. The other is that the porosity is high. Understanding the phase and porosity transformation mechanism of NiTi alloys during vacuum sintering is necessary to solve both problems.

For purpose of this paper, it was planned to prepare dense NiTi alloy by powder metallurgy technology, and research the transformation of phase and porosity in sinter process. The research results can provide guidance for further optimizing sintering technology.

2. Experiment

Content from this work may be used under the terms of the Creative Commons Attribution 3.0 licence. Any further distribution of this work must maintain attribution to the author(s) and the title of the work, journal citation and DOI.
Published under licence by IOP Publishing Ltd

IOP Conf. Series: Materials Science and Engineering **204** (2017) 012023 doi:10.1088/1757-899X/204/1/012023

2.1 Sample preparation

The nickel powder (particle size 2.45 um, purity> 99.75%) and titanium powder (particle size 5 um, purity> 99.9%) used in the experiment were commercially available. Weigh them according to equal molar ratio, and then put them into vacuum ball-milling pot together with proper amount of anhydrous ethanol, and then wet-grind them in the planetary ball mill for 12 hours, with milling speed of 340r / min, and ball-to-powder ratio of 15: 1. After finishing the ball milling, we got mixed powder through drying, sieving and prilling.

The green compact was got at the pressing force of 10 ton by four-pillar hydraulic machine. The size of the green compact is 16×16×5 mm. Then put them into vacuum furnace for sintering, at sintering temperature of 900℃ and 1050℃ respectively, for 2h, 4h, 6h and 8h each with the heating rate of 5℃ / min. After sintering, samples were cooled by furnace cooling.

2.2 Sample test

The X-ray diffraction (XRD) analysis was carried out for phase identification. Microstructure was investigated by SEM in a back-scattered electron (BSE) mode. Archimedes method was used to test the density of sample. Porosity was calculated by formula (1)[12].

$$\theta = (1 - \frac{\rho}{\rho_s}) \times 100\% \quad (1)$$

Where ρ is density of the samples, $g.cm^{-3}$; ρ_s is theoretical density of NiTi alloy, $\rho_s=6.45g.cm^{-3}$.

3. Results and Discussion

3.1 Phase analysis

Figure 1 is the XRD testing results of NiTi alloy specimens prepared with different sintering time at 900℃. As can be seen from the figure, in addition to B2-NiTi and B19 '- NiTi phases, there are also a lot of $NiTi_2$ and Ni_3Ti impurity phases and a little of Ni and Ti elements in the sample sintered for 2 hours. The diffraction peak intensity of $NiTi_2$ and Ni_3Ti impurity phases sharply decreases with the increase of sintering time, and become very weak after 6 hours sintering time. $NiTi_2$ and Ni_3Ti diffraction peak still exists when increase the sintering time to eight hours. And compared with the sample sintered for 6 hours, there is no further weaken.

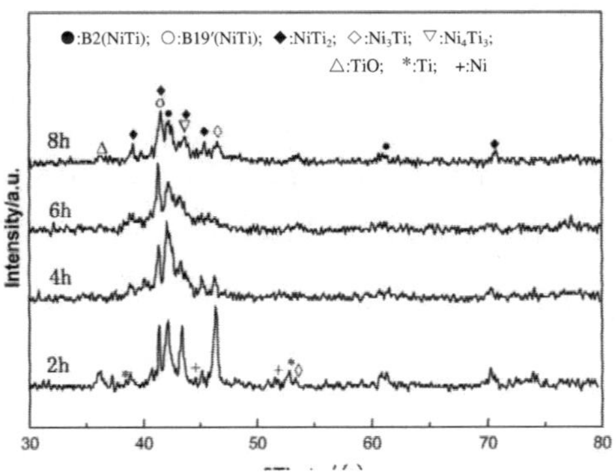

Figure 1. Typical XRD patterns of the samples sintered at 900℃ with different sintering time

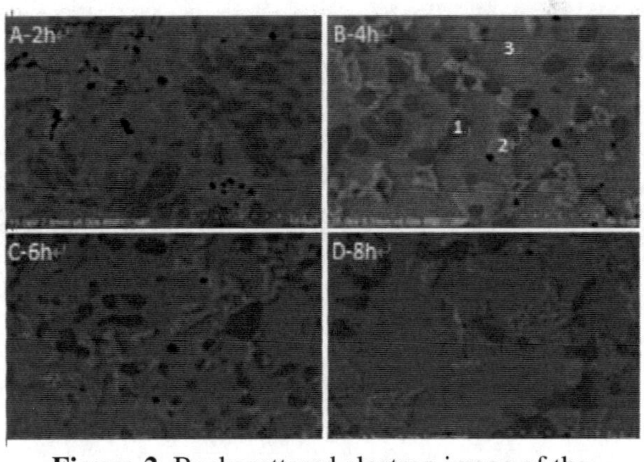

Figure 2. Backscattered electron image of the specimens with different sintering time at 900℃

As can be seen from figure 2, there are three main contrast images in the BSE pictures. The area with different contrast was studied by energy spectrum analysis, and the results are shown in table 1. The dark grey area 1 is $NiTi_2$ phase, the offwhite areas 2 is Ni_3Ti phase, the grey area 3 is NiTi phase. And in the picture, black circular area is the pores in the material. NiTi phase increases gradually with the increase of sintering time, and $NiTi_2$ and Ni_3Ti impurities content gradually reduce. But the content of $NiTi_2$ and

IOP Conf. Series: Materials Science and Engineering **204** (2017) 012023 doi:10.1088/1757-899X/204/1/012023

Ni_3Ti in sample with 6 and 8 hours sintering time are almost same. This shows the $NiTi_2$ and Ni_3Ti impurity phases will not reduce with the increase of sintering time when the sinter time was reduced to a certain degree. So, it does not to completely eliminate the impurity in the organization through long-time sintering. The morphology of $NiTi_2$ in the organization also changed with the increase of sintering time, finally the irregular thick sheet become into small sphere.

Laeng[13]et al. believe that, during the sintering process, NiTi phase is not directly formed from mutual diffusion between Ni and Ti. As the sinter temperature increases, α-Ti is converted to β-Ti[14]at 760-820℃; then Ni diffuses into β-Ti, resulting in the formation of β-Ti(Ni) solid solution; after the β-Ti(Ni) solution is saturated with Ni, reaction between Ni and β-Ti(Ni) begins (β-Ti(Ni)+Ni→$NiTi_2$), followed by the subsequent reaction of $NiTi_2$+Ni→NiTi if the thermodynamics condition allows. At the same time, Ti diffuses into Ni, resulting in a solid solution of Ni(Ti); the continued diffusion of Ti converts the saturated Ni(Ti) solid solution into Ni_3Ti phase, and further leads to the conversion of Ni_3Ti+Ti→NiTi. The above diffusion reaction needs time. Large amount of $NiTi_2$ phase and Ni_3Ti phase can be detected in a test sample after only 2-hour sintering at 900℃. It due to the fact that those initially formed $NiTi_2$ phase and Ni_3Ti phase fail to be completely converted into NiTi phase through the secondary reactions as the sintering time is short. With the increase of sintering time, more and more $NiTi_2$ and Ni_3Ti turned into NiTi phase, which made the content of $NiTi_2$ and Ni_3Ti in tissue reduce, so diffraction peak intensity of $NiTi_2$ and Ni_3Ti impurity phase became weak gradually. However, according to the mutual diffusion mechanism between Ni and Ti, the main reason that $NiTi_2$ and Ni_3Ti in tissue reduced was that $NiTi_2$ and Ni_3Ti respectively occurred reaction with Ni and Ti, which transfer into NiTi phase. With the process of transformation, the number of Ni and Ti became less and less, when simple substance of Ni and Ti ran out, the residual $NiTi_2$ and Ni_3Ti would be remained in tissue, which would not decrease with the extension of sintering time. So even sintering the sample for 8 hours under 900℃, there was still little $NiTi_2$ and Ni_3Ti impurity phase in tissue.

Table1. EDS analysis of phases identified in Fig. 2

	Ni(at%)	Ti(at%)	phase
charcoal grey 1	34.39	65.61	$NiTi_2$
offwhite 2	74.60	26.40	Ni_3Ti
grey 3	48.48	51.52	NiTi

Figure 3 were XRD test results of samples which were sintered for 2 hours under 900℃and 1050℃. It could be found that the diffraction peak intensity of NiTi2 and Ni3Ti of the sample sintered at 1050℃ was weaker. It means that the impurity content of this sample is less. So under the same sintering times, improving sintering temperature could obviously promote the sintering reaction, made the content of $NiTi_2$ and Ni_3Ti impurity phase reduce, and increased the content of B2 and B19' phase.

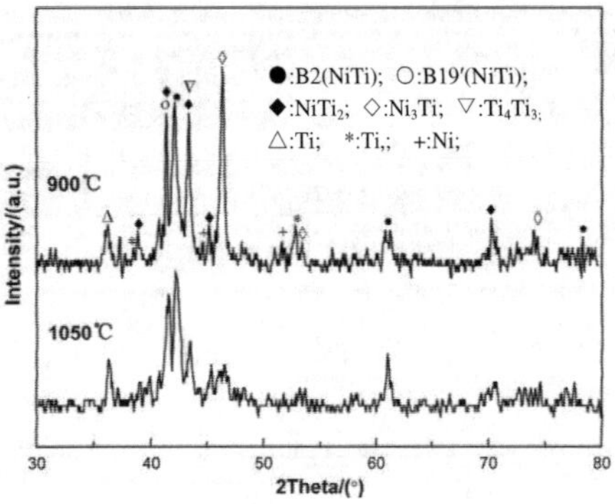

Figure 3. XRD pattern of specimens sintered 2 hours at different sintering temperature

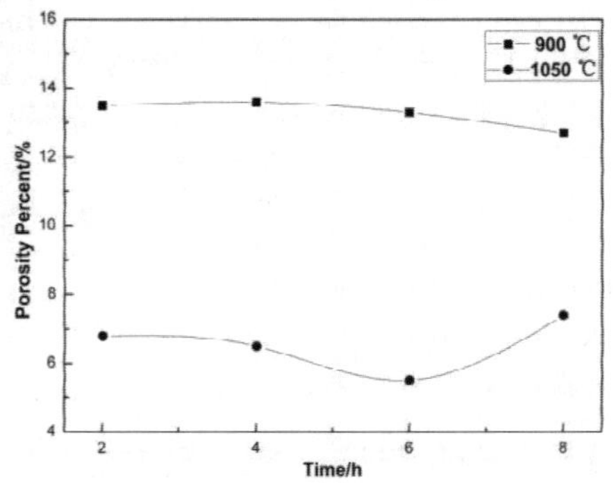

Figure 4. Effect of sintering temperature and time on porosity of NiTi alloy specimens

A transformation of β(Ti)+$NiTi_2$→L (liquid phase) occurs when the temperature reaches 942℃. As the temperature increases to 984℃, residual NiTi2 will be transformed to NiTi+L (liquid phase). Both the generated liquid phase and higher sintering temperature increase diffusion rates of Ti and Ni atom, which facilitates sintering reaction and allows more $NiTi_2$ and Ni_3Ti to be transformed into NiTi phase. In addition, thermal explosion reaction occurs when the generated titanium-rich liquid phase contacts the nickel-rich zone, and NiTi phase is formed[14]. Therefore, the diffracted intensity of $NiTi_2$ and Ni_3Ti impurity phase achieved in the sample tissue after sintering under 1050℃ for 2 hours is weaker (see Fig. 3), and it means that the content of the impurity phase is less.

It is seen from experimental results that sinter time and sinter temperature has relative evident influence on phase composition of samples. Expanding sinter time and increasing sinter temperature both reduce percentages of impurity phases $NiTi_2$ and Ni_3Ti in sample tissues. Compared to the high sintering temperature, longer sintering time is required for the same outcome at low sintering temperature.

3.2 Porosity analysis

The density of the sintered sample was measured based onArchimedes drainage method. Porosity of the sintered sample was calculated by formula (1). The results are demonstrated in Figure 4. It could be observed that the porosity of the samples sintered at 900℃ decreased with the extension of sintering time; nevertheless, the decrease extent was small. The porosity of the samples sintered at 900℃ ranged from 12.7% to 13.6%. Their density ranged from 5.58g.cm^{-3} to 5.63g.cm^{-3}. The theoretical density of the compact NiTi alloy is 6.45g.cm^{-3}. When the sintering temperature was increased to 1050℃, the porosity of the sintered sample decreased at the beginning while increased later with the increase of sinter time. When the sintering time was 6 hours, the density of the sample reached up to the maximum value of 6.10g.cm^{-3}; at this time, the porosity was 5.43%.

Extension of sintering time may lead to two phenomena which will then cause a decrease of alloy's porosity. (1) With extension of sintering time, diffusion between Ti and Ni becomes more sufficient while combination between particles becomes closer, which will make part of original pores diminish o rdisappear [15]; (2) With extension of sintering time, crystal grains will gradually grow bigger. During movement of the grain boundary, a large number of pores which have been swept by the grain boundary will disappear, and porosity will gradually decrease[16]. But when sintering time is prolonged to 8 hours,porosity in the tissue will instead increase. Whitney[14] et al. explained that: when the sintering temperature rises to 942℃, transformation of β(Ti)+$NiTi_2$→L will occur and thus produce Ti-rich liquid phase; when the sintering time is long enough, a large amount of liquid can flow to the Ni-rich area through pores by crystallization and react with it to form intermetallic compound, so that pores will be formed at the original location of liquid phase.

4. Conclusions

Phase and porosity transformation characteristics of NiTi shape memory alloys in the process of vacuum sintering were investigated, the conclusions drawn from this study can be summarized as follows:

1) With increase of sintering time, the content of Ni_3Ti and $NiTi_2$ impurity phases decreases, and that of B2 and B19′ increases. Transformation from Ni_3Ti and $NiTi_2$ impurity phases to NiTi can be promoted by raising the sintering temperature.

2) The impurity phases of Ni_3Ti and $NiTi_2$ cannot be thoroughly eliminated by long-time sintering.

3) With increase of sintering time, the porosity of specimens sintering at 900℃ decreases slightly with increase of sintering time. But the porosity of specimen sintering at 1050℃ decreased firstly and then increased. The reason is that the generation mechanism of NiTi phase change when the sintering temperature reach or exceed 942℃.

Acknowledgments

This research is funded by Chongqing Research Program of Basic Research and Frontier Technology（No. cstc2014jcyjA50010）

References

IOP Conf. Series: Materials Science and Engineering **204** (2017) 012023 doi:10.1088/1757-899X/204/1/012023

[1] Fu Tao,Li Hongwei,Wu Feng,Li Wen and Sun Jianmin 2016 *Rare Metal Materials and Engineering***45**pp 1128-1131
[2] Kong Xiangque,Jin Xuejun and Liu Jiannan2016Journal of Functional Material**47**1007-1011
[3] BingHui Dong,Feng Wu,Zafer Alajmi,Chen Zhang,Tao Fu and Yuan Ge2014*Rare Met.***33** pp 21–27
[4] Zhou Chao,He Zhirong,Liu Lin and Li Xudong 2015*Hot Working Technology***44**pp 7-11
[5] Song Jie,Yin Hesheng,Xu Chengkai,Li Xiaogao,Song Xianyi and Kong Weichuan 2014 *Development & Innovation of Machinery & Electrical Products***27**pp 23-25
[6] H.J. Jiang, S. Cao, C.B. Ke, X. Ma and X.P. Zhang2013*Materials Letters***100**pp 74–77
[7] Gang Chen,Klaus-Dieter Liss and Peng CA2015*METALLURGICAL AND MATERIALS TRANSACTIONS A***46**pp 5887-5899
[8] Kang Xiaoyu,Li Yanfeng,Yin Xiangqian,Xie Haofeng and Mi Xunjun 2013*Heat Treatment of Metals***38**pp 23-26
[9] C. Bewerse,L.C. Brinson and D.C. Dunand2016*Acta Materialia***115**pp 83-93
[10] Liu A L,Gao Z Y,Gao L, Cai W and Wu Y2007*Journal of Alloys and Compounds***437**pp 339-343
[11] Zhao Tengfei,Lu Xin and Qu Xuanhui2012*Powder Metallurgy Technology***30**pp 300-306
[12] Xu Wei,Shi Xinying,Guan Ruifeng, Zhang Xinping and Zhao Ke 2011*Journal of Medical Biomechanics***26**pp349-354
[13] Jamaluddin Laeng,Zhimeng Xiu,Xiaoxue Xu,Sun Xudong,Ru Hongquiang,Liu Yinong2007 *Physica Scripta***129**pp250-254
[14] M.Whitney,S.F.Corbin and R.B. Gorbet2008*Acta Materilia***56**pp 559-570
[15] Ma Xuliang,Li Li,Zhu Chengwu,Wang Xiang and Zheng Yufeng 2011*Journal of Materials Engineering***3**pp 6-10
[16] Li Mingwei,Yan Yinbiao and Cheng Min2011*Powder Metallurgy Technology***29**pp 372-379

IOP Conf. Series: Materials Science and Engineering **204** (2017) 012024 doi:10.1088/1757-899X/204/1/012024

Rigid palm oil-based polyurethane foam reinforced with diamine-modified montmorillonite nanoclay

MohdHaziqDzulkifli[1] , MohdYazidYahya[1] andRohah A. Majid[2]

[1]Center for Composites (CfC), Institute for Vehicle Systems Engineering (IVeSE), University Teknologi Malaysia (UTM), 81310 Johor Bahru, Johor, Malaysia
[2]Department of Polymer and Bioprocess Engineering, Faculty of Chemical and Energy Engineering, University Teknologi Malaysia (UTM), 81310 Johor Bahru, Johor, Malaysia

E-mail: mohadz90@yahoo.com,yazid@fkm.utm.my,rohah@cheme.utm.my

Abstract: This paper presents work on organically-modified montmorillonite (MMT) nanoclay embedded in rigid palm oil-based polyurethane (PU) foam. MMT was modified with organic surfactant diamino propane (DAP). PU foam was fabricated in closed mold, and the amount of DAP-MMT was varied in each foam formulation. The obtained foam was tested for its microstructure and morphology. Appearance of peaks from infra-red spectra corresponding to N-H, C=O, and C-N confirms the formation of PU networks. Scanning electron microscopy (SEM) revealed fine, closed-cellular structure at low clay loading; increasing DAP-MMT content induced larger cell sizes with blowholes. X-ray diffraction (XRD) indicates fully-exfoliated clays at 1 wt. % and partial-exfoliation at 3 wt. % clay loading, suggesting clumping of clays as DAP-MMT content increased.

1. Introduction

Due to its high surface area-to-volume ratio, nano-reinforcements such as nano carbon fibers (CNFs), nano carbon tubes (CNTs), graphenes, and nanosilicas can impart extreme enhancement in mechanical properties when embedded even with little amount in polymers such as polypropylene (PP), polyurethane (PU), and epoxy resin [1–3], which might be the reason they are widely utilized in industries. Even so, the aforementioned materials are synthetically-manufactured; and as such are not environmental friendly and often cost consuming. Researchers over recent years are therefore searching for a more renewable alternatives; and among the 'green' nano-reinforcements are montmorillonite (MMT) nanoclay.

Montmorillonites are smectite clays which are well-known for its high swelling capacity and good adsorption/absorption properties [4]. Often these nanoclays undergo surface modification to increase its compatibility with polymer matrix, due to their difference in polarities. According to Zeng et al [3], ion exchange method is the most common method in clay surface treatment, enabling larger interlayer spacing or d-spacing within the nanoclay. This in return allows the nanoclays to be exfoliated or efficiently dispersed within the polymer matrix. Utilization of numerous surfactants for ion exchange method with MMT have been reported elsewhere [5,6]; in this research, we have chosen diamino propane (DAP) as the organic modifier. DAP is a linear-typed surfactant with two amine groups at each end of it polymer chain.

Among the common polymers embedding organoclays in its system is polyurethane (PU). Exists in many forms – foam, thin film, coating, elastomer, etc. – they are vastly utilized in industries as insulation, in deepwater pipeline, packaging, damper, and even for automotive and aeronautic

Content from this work may be used under the terms of the Creative Commons Attribution 3.0 licence. Any further distribution of this work must maintain attribution to the author(s) and the title of the work, journal citation and DOI.
Published under licence by IOP Publishing Ltd

applications [7–9]. However, PUs consume petroleum feedstock in its production; and this is not favorable with today's depleting petroleum supply and sustainability issue. Efforts were made to find PU's greener counterpart, and it has been proven over the past years that oil from plants such as rapeseed, soybean, castor, and palm oil can be modified and used in production of PU [10–13]. This research will focus on palm oil-based PU foam since palm oils are abundant in South East Asia countries like Malaysia and Indonesia.

Current work will be on rigid, palm oil-based PU foam embedded with DAP–modified MMT. The foam was fabricated in closed mold; and the amount of DAP-MMT was varied in each foam formulation. The obtained foam was then tested for its morphology and microstructure. Current work is a continuity of our previous work, aimed to clarify past made claims and to further understand the effect of organic modifier on nanoclay and its polymeric system.

2. Experimental

2.1 Materials

The materials used in this study are MMT K10 nanoclay; hydrochloric acid, HCl; diamino propane, DAP (Sigma-Aldrich); sodium carbonate Na_2CO_3 (BioBasic Inc.); 4,4'-diphenylmethane diisocyanate, *p*-MDI (Merck Sdn. Bhd.); polyester palm oil-based polyol, POP (MaskimiPolyol, Malaysia); silicon surfactant Tegostab B8404 (Evonik (M) Sdn. Bhd.); and distilled water.

2.2 Modification of MMT

Pristine MMT K10 was first dispersed in distilled water along with Na2CO3 and stirred at 250 rpm for approximately 24 hours. The mixture was then filtered and washed several times prior to dried in vacuum oven at 60 °C for 3 to 4 hours. The obtained dried clay was ground into fine powder and re-immersed in distilled water and stirred at 250 rpm for approximately 24 hours. Later, the mixture was heated to approximately 70 °C, before dropwise addition of HCl and DAP aqueous solution. The stirring continues for another 3 hours at 250 rpm and 70 °C, before the mixture was filtered and the filtrate washed several times. It is then dried in vacuum oven at 60 °C for 3 to 4 hours, and the dried clay was ground into fine powder for next stage.

2.3 Fabrication of PU nanocomposites foam

For this study, POP:p-MDI ratio was set at 1:1. Firstly, POP was mixed with distilled water, silicon surfactant, and DAP-MMT (1 to 4 wt. %), stirred at 200 rpm for about 10 minutes, forming homogenous mixture called 'premix'. *p*-MDI was then added to the premix and stirred at 500 rpm for 10 to 15 seconds, before poured into closed aluminium mold. It is then left for foaming and curing for 24 hours at room temperature before being demolded, and prepared for testing.

2.4 Fourier-transform infra-red (FTIR) spectroscopy

Infra-red spectroscopy via attenuated total reflectance (ATR) sampling was carried out using Perkin-Elmer Frontier. The foam was scraped into powder and placed onto the ATR crystal. The tip of the force gauge was adjusted to slightly touch the sample. Analysis was then carried out at wavelength range of 4000 cm^{-1} to 650 cm^{-1}, resolution of 4.0 cm^{-1}, and 16 total scans.

2.5 X-Ray diffraction (XRD)

XRD was carried with 2θ range of 2 to 10°. The value of d-spacing of the organoclay was determined using Bragg's equation, as given by Eq. 1:

$$\lambda = 2d \sin \theta \qquad (1)$$

whereλ is the X-ray radiation wavelength used (λ=1.54056 Å), *d* is the interlayer spacing of the MMT nanoclay, and θ is the measured diffraction angle.

2.6 Scanning electron micoscopy (SEM)

SEM JEOL JSM 6390-LV was used to study the surface morphology of the foam. Samples of size 5 x 5 x 5 mm^3 were prepared and coated with gold. The opening voltage was set at 10 kV.

3. Results and Discussion

The infrared spectra of pristine MMT and modified MMT are presented in Figure 1(a), whereas Figure 1(b) illustrates the infrared spectra of pristine PU, and with added 1 and 3 percent weight (wt. %), respectively. Both infra-red spectra for pristine MMT and DAP-MMT portrays the basic characteristic bands of nanoclays: ~3500 cm^{-1} (OH stretching), 1029 cm^{-1} (Si-O stretching), and 797 cm^{-1} (Al-O bending) [14]. Modifying MMT with DAP, several new peaks were present in the spectrum; the most prominent was at 3463 cm^{-1}, denoting successful inclusion of DAP into MMT gallery. Further supporting is presence of bands at 3043 and 2948 cm^{-1}, attributed to the absorption band of DAP symmetric and asymmetric C-H, respectively. Other bands include 1561 cm^{-1} (NH_2 stretching), 1457 cm^{-1} (CH_3 and CH_2 asymmetrical stretching), and 1325 cm^{-1} (CH_3 and CH_2 symmetrical stretching) [15]. As for pristine and nanocomposite foam spectra, presence of peaks at 1712 cm^{-1}, 1509 cm^{-1}, and 1216 cm^{-1} confirms the formation of PU network as these bands represent the bending vibration C=O, N-H, and C-N, respectively. Band at 3314 cm^{-1} and 2280 cm^{-1} represent the free hydroxyl, –OH and isocyanates, –NCO groups in the system, respectively. Peaks at 2924 cm^{-1} and 2850 cm^{-1} were ascribed to the stretching vibration of C-H asymmetric and symmetric, respectively [13].

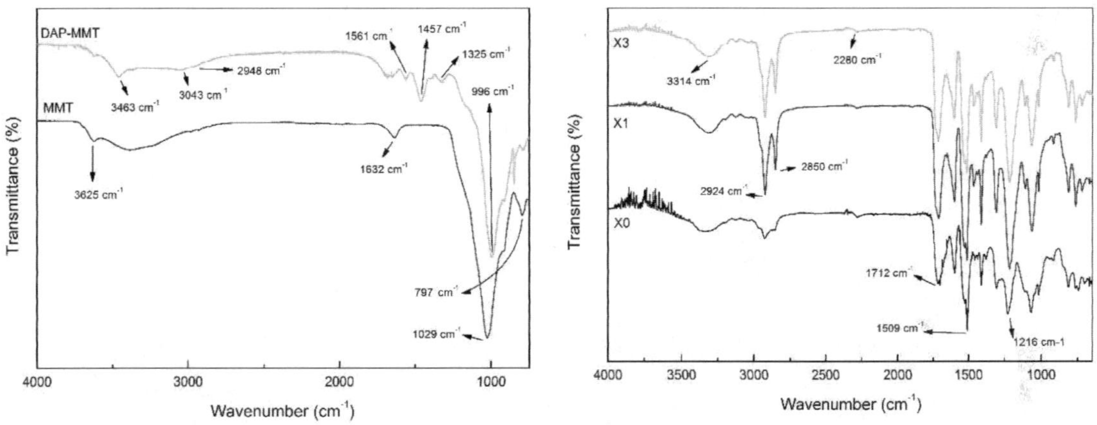

(a) **(b)**

Figure 1. Infra-red spectra of (a) pristine and organically modified MMT; and (b) pristine PU foam (X0) and PU foam with 1 and 3 wt. % DAP-MMT (X1 and X3)

Figure 2(a) shows the XRD curves for pristine MMT and DAP-MMT, whereas XRD curves for PU foam with 1 wt. % and 3 wt. % DAP-MMT are shown in Figure 2(b). For the MMT's curves, it can be obviously seen the shifting of peaks from 8.95° for pristine MMT towards the left side at 8.05° for DAP-MMT. This signifies larger interlayer spacing or d-spacing was achieved as the diffraction angle for pristine MMT corresponded to d-spacing of 9.872 Å, whereas DAP-MMT 10.974 Å; confirming successful inclusion of DAP into MMT gallery [14]. As for XRD curves for nanocomposite foams, no significant peaks can be detected upon addition of 1 wt. % DAP-MMT (sample X1), indicating the clays were fully-exfoliated inside the PU [5,16]. Increasing the content to 3 wt. % (sample X3), however; slightly broad elbow can be detected at 2θ value of 6.16°, corresponding to d-spacing value of 14.34 Å. This finding might suggest that only partial-exfoliation clay dispersion can be achieved at higher values of clay loading, in agreement with past publication [17]. As we discussed in our previous work, we thought that the clays tend to agglomerate at higher clay loadings to justify the drop in mechanical properties; and this finding supports our past claims. Past literatures also reported clay agglomeration at higher clay loading [18,19].

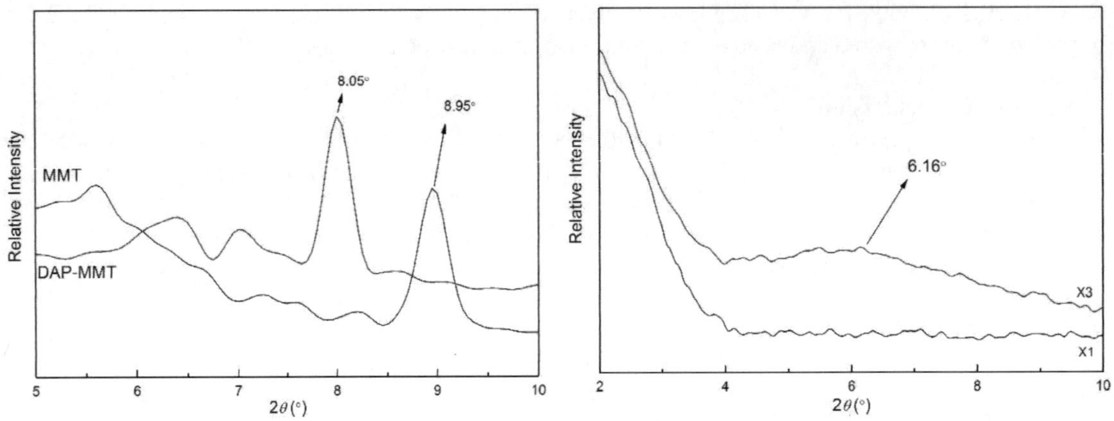

(a) **(b)**

Figure 2. XRD curves of (a) pristine MMT and DAP-MMT; and (b) PU foams with 1 wt. % and 3 wt. % DAP-MMT

The microscopic images of PU foams with 0 wt. %, 1 wt. %, and 3 wt. % DAP-MMT are presented in Figures 3(a) through (c), respectively. All obtained foams are of closed-cell foams. It can be observed that the cell sizes were reduced upon addition of 1 wt. % DAP-MMT into the foam formulation. This is thought due to the exfoliated structure, as obtained in XRD result. As clays can function as bubble nucleation site, efficient dispersion of DAP-MMT inside the PU foam matrix can contribute in preventing coalescence, and producing smaller, fine call structure [20]. As the content was increased to 3 wt. %, the cell sizes increase again. As we claimed previously, it is thought that clay agglomeration occurred at this point; and clay clumping tends to cause cell coalescence by 'over-blowing' carbon dioxide gas CO2 inside the cells, hence increasing their cell diameter [21,22]. This finding supports our past claim in explaining the reduction in mechanical properties of the PU nanocomposite foams at higher DAP-MMT loading.

Figure 3. SEM images of (a) pristine PU foam; (b) PU foam with 1 wt. % DAP-MMT; and (c) PU foam with 3 wt. % DAP-MMT

4. Conclusion

In current work, pristine montmorillonite (MMT) was successfully modified with diamino propane (DAP) and embedded inside rigid, palm oil-based polyurethane (PU) foam to study its effect on the foam's structure and morphology. From the Fourier-transform infrared (FTIR) spectroscopy, the tethering of DAP into MMT gallery was confirmed via detection of peak at 3463 cm^{-1}. Peaks that corresponds to C=O, N-H, and C-N bands also confirmed the formation of PU network. Successful inclusion of DAP into MMT was also confirmed through the peak-shifting X-ray diffraction (XRD) pattern, resulting in expanded interlayer gallery from 9.872 Å to 10.974 Å. XRD results also revealed that fully-exfoliated microstructure was achieved for foams with 1 wt. % DAP-MMT loading, and partial-exfoliation at 3 wt. %. Through scanning electron microscopy (SEM), it was noted the reduction in cell size upon introduction of 1 wt. % DAP-MMT into the system, and increment in cell size at 3 wt. %; which we believe owed to the dispersion of the DAP-MMT into the system. To conclude, inclusion of organically-modified nanoclay into PU foam system results in change in morphology and microstructure of the foam.

Acknowledgement
The authors would like to thank Zamalah UTM for providing scholarship to the author, and also to Ministry of Higher Education under Fundamental Research Grant (FRGS) R.J130000.7844.4F235.

5. References

[1] Tjong S C 2006 *Mater. Sci. Eng.***53** p 73
[2] Balakrishnan H, Ibrahim M, Wahit M U and Hassan A 2011*Polym. Compos.***32** p 1927
[3] Zeng Q H, Yu A B, Lu G Q M and Paul D R2005 *J. Nanosci. Nanotechnol.***5** p 1574
[4] Xi Y, Frost R L and He H2007 *J. Colloid Interface Sci.* **305** p 150
[5] Palanisamy A 2013*Polym Compos.***34** p 1306
[6] Salahuddin N, El-enein S A A, SelimA and El-dien O S 2010*Appl. Clay Sci.***47** p 242
[7] Chian K S and Gan L H1998 *J. Appl. Polym. Sci.***68** p 509
[8] Rashmi B J, Rusu D, Prashantha K, Lacrampe M F and Krawczak P 2013 *J. Appl. Polym. Sci.***128** p 292
[9] Prociak A, Rojek P and Pawlik H2012 *J. Cell. Plast.***48** p 489
[10] Kurańska M, Cabulis U, Auguścik M, Prociak A, Ryszkowska J and Kirpluks M2015 *Polym. Degrad. Stab.***127** p 11
[11] Guo A, Javni I and Petrovic Z 2000*J. Appl. Polym. Sci.***77** p 467
[12] Stirna U, Lazdiņa B, Vilsone D, Lopez M J, Vargas-Garcia M d C, Suárez-Estrella F and Moreno J 2012*J. Cell. Plast.***48** p 476
[13] Dzulkifli M H, Yahya M Y, MajidR A and Akhir F S M 2014*JurnalTeknologi.***68** p 53
[14] Zaini M, Majid R A and Nikbakht H 2015*Appl. Mech. Mat.***695** p 224
[15] Emara A A A, Saleh A A and Adly O M I 2007*Spectrochim. Acta Part A.***68** p 592
[16] Chang J–H and An Y U 2002*J. Polym. Sci. Part B Polym. Phys.***40** p 670
[17] Liang K and Shi S Q 2011*J. Appl. Polym. Sci.***119** p 1857
[18] Mohan T P and Kanny K 2011*J. Reinf. Plast. Compos.***30** p 152
[19] Xiong J, Zheng Z, Jiang H, Ye S and Wang X 2007 *Compos. Part A Appl. Sci. Manuf.***38** p 132
[20] Lee L J, Zeng C, Cao X, Han X, ShenJ and Xu G 2005 *Compos. Sci. Technol.***65** p 2344
[21] Xu Z, Tang X, Gu A and Fang Z2007 *J. Appl. Polym. Sci.***106** p 439
[22] Harikrishnan G, Patro T U and Khakhar D V 2006*Ind. Eng. Chem. Res.***45** p 7126

IOP Conf. Series: Materials Science and Engineering **204** (2017) 012025 doi:10.1088/1757-899X/204/1/012025

The Effect of Various Acids on Properties of Microcrystalline Cellulose (MCC) Extracted from Rice Husk (RH)

A S Nur Hanani[1], A Zuliahani[1], W I Nawawi[1], N Razif[1] and A R Rozyanty[2]

[1]Faculty of Applied Sciences, UniversitiTeknologi Mara (UiTM) Perlis, Malaysia
[2]School of Material Engineering, Universiti Malaysia Perlis, Malaysia

E-mail: nurhananiabuseman92@gmail.com,zuliahani@perlis.uitm.edu.my, wi_nawa wi@perlis.uitm.edu.my, razifmn@perlis.uitm.edu.my, rozyanty@unimap.edu.my

Abstract. Microcrystalline cellulose (MCC) was successfully extracted from rice husk (RH) via acid hydrolysis process using nitric acid (HNO_3) in comparison with sulphuric acid (H_2SO_4) and hydrochloric acid (HCl). MCC-RH extracted using HNO_3 produced the highest percentage yield at 83.5% as compared to H_2SO_4 and HCl at 80.6% and 81.8% respectively. Analysis of Fourier Transform Infrared (FTIR) spectroscopy affirmed the successive elimination of non-cellulosic material from RH cellulose resulting highly purified MCC-RH. X-ray Diffraction (XRD) analysis showed MCC-RH treated with HCl gives the highest crystallinity index value of 54.2% while HNO_3 and H_2SO_4 produced comparable results of 52.4% and 49.7% respectively. The results indicate successive extraction of MCC-RH using HNO_3 that has great potential to replace strong acid such as H_2SO_4 and HCl in acid hydrolysis.

1. Introduction

MCC is a fine, white, odorless, crystalline powder and biodegradable material which composed of porous particles. It can be isolated from any material that contained high cellulose ranging from pure cellulose, commercial grade cellulose to lignocellulocis materials [1]. MCC is applied in pharmaceutical, cosmetics, food industries and acts as a water retainer, suspension stabilizer and reinforcement filler in composite preparation [2].

Generally, MCC is isolated from wood pulp and purified cotton linters via acid hydrolysis process [3]. Acid hydrolysis is commonly used to isolate MCC whereby the amorphous region of cellulose was disintegrated resulting a highly crystalline material with different degree of crystallinity such as MCC and nanocrytalline cellulose (NCC) [4]. Previous studies have been used strong acids such as H_2SO_4 and HCl to eliminate amorphous region during acid hydrolysis [1,2]. Lignin and hemicellulose were removed from RH undergone pre-treatments. Alkaline treatment was performed to solubilize hemicellulose while bleaching treatment was used to remove lignin [5].

Recent studies reported that MCC also can be obtained from the agricultural wastes such as coconut husk fibres, groundnut husk, rice straw and rice husk [6-9]. RH is an agricultural waste and a huge quantity of RH has been generated during the rice milling process. It was used as fertilizer, building material, heat insulator and produced RH ash for the production of silica [10]. However, only small amount of RH was recycled, whilst large quantities of RH are burn in open air which caused air pollution and resulted damage to the land.

RH is composed of 25-35% cellulose, 26-31% lignin and 18-21% hemicellulose [11]. In the rice husks cell wall, cellulose exists in a complex lignocellulosic matrix, surrounded by lignin and hemicellulose [12]. Since RH contains high composition of cellulose, it has great potential for producing MCC via acid hydrolysis process.

Content from this work may be used under the terms of the Creative Commons Attribution 3.0 licence. Any further distribution of this work must maintain attribution to the author(s) and the title of the work, journal citation and DOI.
Published under licence by IOP Publishing Ltd

In this study, HNO_3 was used as an alternative to replace the usage of strong acids in the MCC isolation due to its miscibility in water and less corrosive characteristic. The significant of this study is to encourage the utilization of MCC isolated from RH via acid hydrolysis using HNO_3 in comparison with H_2SO_4 and HCl.

2. Materials

RH was collected from BERNAS, a local rice mill situated in Pendang, Kedah, Malaysia. Sodium hypochlorite (NaOCl) and sodium hydroxide (NaOH) were used as bleaching agents while HNO_3, H_2SO_4 and HCl were used for acid hydrolysis process. All chemical were purchased from Fluka and were used without further purification.

3. Experimental Method

3.1. Alkaline and bleaching treatments of RH

The alkaline treatment was performed to solubilize hemicellulose. 10g of RH was reflux with 120 mL of 1M NaOH at 80 °C for 1 h and 30 min, and then washed with distilled water for removal of solubilized components until pH 7 was achieved. The alkaline treated RH then was undergone bleaching treatment and refluxed with 140 mL of 5% NaOCl at 80 °C for 18 min [13,14]. The cellulose obtained from the treated RH was filtered and rinsed several times using distilled water. It was dried in an oven at 70 °C until constant weight was achieved and the weight of the cellulose was recorded.

3.2. Acid hydrolysis

The obtained cellulose was hydrolyzed using 140 mL of 0.5M HNO_3 at a temperature of 80 °C for 30 min under continuous stirring. The MCC collected was filtered and rinsed with distilled water until pH 7 was achieved. It was then oven dried at 70 °C for 24 h. After constant weight was achieved, the weight of the MCC was recorded and it was grounded into fine powder using rotary ball mill. The same steps were repeated for 0.5M H_2SO_4 and 0.5M HCl [14].

3.3. Percentage yield

The percentage yield of MCC was calculated using the formula computed by recent studies [15].

$$\text{Percent yield of MCC (\%)} = \frac{A}{B}\, x\ 100\% \qquad (1)$$

where,

A = weight of obtained MCC
B = weight of cellulose

3.4. Fourier Transform Insfrared (FTIR) spectroscopy

FTIR analysis of untreated RH and the obtained MCC was performed using Perkin Elmer FTIR spectrometer 1650 within the wavenumber range of 4000 cm^{-1} to 600 cm^{-1}. FTIR spectra of the samples were recorded using Nicolet's Avatar 360 at 32 scan with the spectra resolution of 4 cm^{-1}.

3.5. X-Ray Diffraction (XRD)

X-ray diffraction was performed to determine crystallinity index (CrI) of the obtained MCC using different acid used. Diffraction patterns were obtained using a PANalyticalX'PertPRO Multi-Purpose Diffractometer with Cu Kα radiation. CrI were calculated via:

$$CrI = \left(\frac{I_{002} - I_{am}}{I_{002}}\right) \qquad (2)$$

where,

I_{002} = intensity of the 002 peak at about $2\theta = 22°$ and I_{am} is the intensity corresponds to the peak at about $2\theta = 18°$

4. Results and Discussions

4.1. Percentage yield of MCC

The percentage yield of MCC during acid hydrolysis process was determined and the data was summarized in Table 1. Different type of acid with same molarity was used in the isolation process of MCC from RH cellulose. The different acids used were HNO_3, H_2SO_4 and HCl with 0.5M concentration.

Table 1. Percentage Yield of MCC Isolated by Different Acids Used

MCC Samples	Percentage Yield [%]
0.5M HNO_3	83.5
0.5M H_2SO_4	80.6
0.5M HCl	81.8

Table 1 shows the highest percentage yield of MCC is obtained by 0.5M HNO_3 which produced 83.5% MCC. The MCC yield acquired from 0.5M H_2SO_4 and 0.5M HCl showed comparable result with 80.6% and 81.8% respectively. This is due to the ability HNO_3 to solubilize more amorphous region of cellulose that contribute to increase the percentage of MCC yield. Hydronium ions in mineral acids able to penetrate and remove the excessive amorphous regions of cellulose, resulting the increase in percentage yield of MCC [16]. In addition, the different in geographical conditions, climate, type of paddy used, sample preparation and method of analysis caused the variation of MCC yield [17].

4.2. Fourier Transform Infrared (FTIR) spectroscopy

Figure 1 shows the FTIR spectra of untreated RH, MCC-RH treated with 0.5M HNO_3, MCC-RH treated with 0.5M H_2SO_4 and MCC-RH treated with 0.5M HCl. For untreated RH, a broad absorption band appears at 3352 cm^{-1} is due to the stretching of –OH groups while the band at 2971 cm^{-1} corresponds to the C-H stretching vibrations. The band at 1636 cm^{-1} indicates the absorption of water and it is related to the bending modes of water molecules due to a strong interaction between cellulose and water [16,18]. All MCC-RH samples appeared at the same absorbance regions as untreated RH that represents –OH groups, C-H stretching and water absorption. Hence, it is affirmed that cellulose components are present in the MCC-RH samples and still remain even after being treated with several chemical treatments.

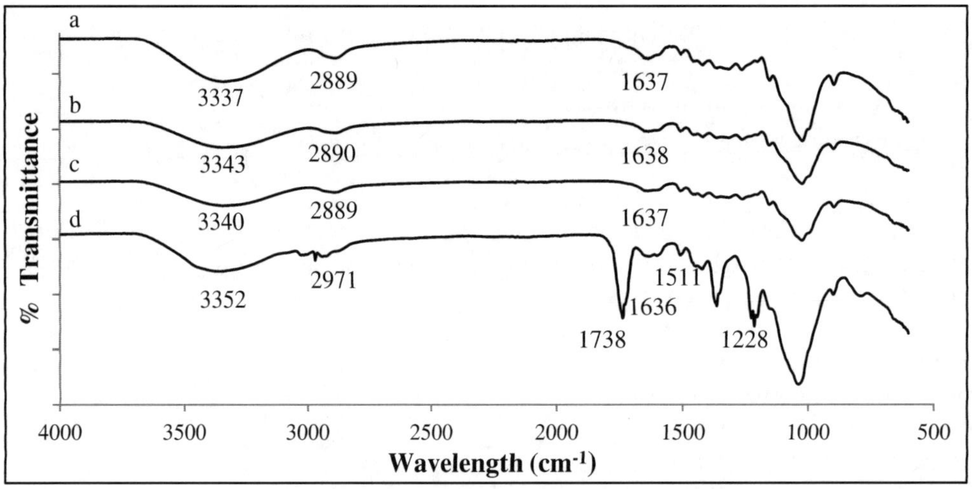

Figure 1. FTIR spectra of (a) MCC-RH treated with 0.5M HNO_3 (b) MCC-RH treated with 0.5M H_2SO_4 (c) MCC-RH treated with 0.5M HCl and (d) untreated RH

Besides, the absorption band located at 1738 cm^{-1} of untreated RH corresponds to the uronic ester and acetyl groups of hemicellulose. Based on previous study, this band normally appears at 1700 – 1740

IOP Conf. Series: Materials Science and Engineering **204** (2017) 012025 doi:10.1088/1757-899X/204/1/012025

cm^{-1} [19,20]. The absence of this band in all MCC-RH spectrums prove hemicellulose is successfully removed from RH cellulose during chemical treatments. A weak band at 1511 cm^{-1} and a strong band appeared at 1228 $cm^{-1}$for untreated RH spectrum are due to C=C aromatic skeletal vibration and C-O stretching of ester linkage of lignin. However, there is no absorption band is observed within the range of 1509 – 1609 cm^{-1} in all MCC-RH spectrum. According to previous work, absence of peaks located in range 1509 – 1609 cm^{-1} indicate complete removal of lignin [21]. Hence, lignin is successfully removed from RH cellulose during alkaline and bleaching treatments

4.3. X-Ray Diffraction (XRD)

The diffraction patterns of MCC-RH treated with 0.5M HCl, MCC-RH treated with 0.5M H_2SO_4, MCC-RH treated with 0.5M HNO_3 and untreated RH are presented in Figure 2. They are highly crystalline native cellulose I and no doublet presence at $2\theta = 22°$ [18]. The crystallinity index of the samples were calculated using Eq. 2. The highest crystallinity value is exhibits by 0.5M HCl at 54.2% as compared to HNO_3, H_2SO_4, and untreated RH at 52.4%, 49.7% and 44.1% respectively. The obtained result is in line with study carried out by Johar*et al* [16].

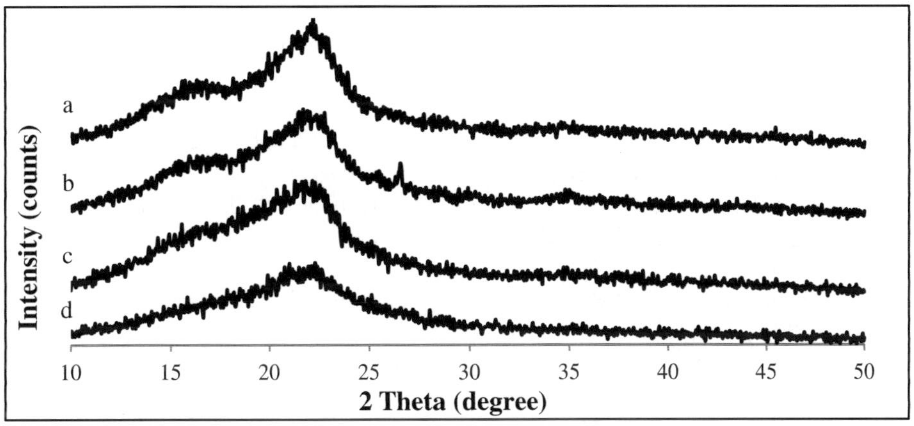

Figure 2. X-ray diffraction patterns of (a) MCC-RH treated with 0.5M HCl, (b) MCC-RH treated with 0.5M H_2SO_4, (c) MCC-RH treated with 0.5M HNO_3 and (d) untreated RH

The increase of crystallinity value is due the acid hydrolysis process that remove the amorphous regions of cellulose [14,18]. During this process, hydronium ions penetrate the accessible amorphous regions of cellulose and allow hydrolytic cleavage of glycosidic bonds which eventually releases individual crystallite [16]. Besides, the increase of cellulose crystallinity were expected to enhance rigidity, stiffness and strength. Thus, the increase of crystallinity value of the treated MCC-RH would be anticipated to provide good physical and mechanical properties of composites.

5. Conclusion

MCC-RH has been successfully extracted from RH using acid hydrolysis process via HNO_3, H_2SO_4 and HCl. The MCC-RH treated with 0.5M HNO_3 shows an optimum result in percentage yield of MCC and it is comparable H_2SO_4 and HCl. By reflecting FTIR results, chemical treatments that have been used in this study such as alkaline, bleaching and acid hydrolysis treatment were successfully removed hemicellulose, lignin and amorphous region from RH cellulose producing a highly crystalline MCC-RH. Whilst, crystallinity value of MCC-RH using HNO_3 is comparable with HCl and H_2SO_4. Thus, HNO_3 has great potential to replace H_2SO_4 and HCl during acid hydrolysis It is also affirmed that HNO_3 able to enhance the properties of MCC-RH in term of yield, purity and crystallinity.

Acknowledgements

Authors wishing to acknowledge Universiti Teknologi Mara (UiTM) Perlis for giving instrumental facilities to complete this research.

References

[1] Ejikeme P M 2008 Investigation of the physicochemical properties of microcrystalline cellulose from agricultural wastes I: orange mesocarp *Cellulose* **15** pp 141-147

[2] Chuayjuljit S, Su-uthai S and Charuchinda S 2010 Poly(vinyl chloride) film filled with microcrystalline cellulose prepared from cotton fabric waste: properties and biodegradability study *Waste Manag. & Res.* **28** pp 109-117

[3] El-Sakhawy M and HassanM L 2007 Physical and mechanical properties of microcrystalline cellulose prepared from agricultural residues *Carbohydr. Polym.* **67** pp 1-10

[4] CostaS S, MorisV A S and RochaS C S2011 Influence of process variables on granulation of microcrystalline cellulose in vibro fluidized bed *Powder Technol.* **207** pp 454-460

[5] TracheaD, DonnotA, KhimecheK, Benelmir R and BrossebN 2014 Physico-chemical properties and thermal stability of microcrystallinecellulose isolated from Alfa fibre *Carbohydr. Polym.* **104** pp 223-230

[6] RosaM F, MedeirosE S, MalmongeJ A, GregorskiK S, WoodD F, MattosoL H C, GlennG, Orts W J and ImamS H 2010 Cellulose nanowhiskers from coconut husk fibers: effect preparation conditions on thermal & morphological behavior *Carbohydr. Polym.* **81** pp 83-92

[7] Ohwoavworhua F O and AdelakunTA 2010 *Indian J. Pharm. Sci.* **72** pp 295-301

[8] IlindraA and DhakeJ D 2008 Microcrystalline cellulose from baggase and rice straw *Indian J. Chem. Tech.* **15** pp 497-499

[9] OhwoavworhuaF O, AdelakunT A and OkhamafeA O 2009 *Int. J. Green Pharm.* **3** pp 97–104

[10] TurmanovaS, SvetlanaG and VlaevL 2012 Obtaining some polymer composites filled with rice husks ash-a review *Int. J. Chem* **4** p 4

[11] LudueñaL, FasceD, Alvarez V Aand StefaniP M 2011 Nanocellulose from rice husk following alkaline treatment to remove silica *Bioresources* **6** pp 1440-1454

[12] MussattoS Iand TeixeiraJ A 2010 *Current Research, Technology and Education, Topics in Applied Microbiology and Microbial Biotechnology* ed A Mendez-Vilas(Spain:Formatex Research Center, Badajoz) pp 897-907

[13] HannaM, BlbyG and MiladinoveV 2001 Production of microcrystalline of rice straw cooking conditions in the sodaethanol-water pulping on the mechanical properties of the produced paper sheet *Biosour. Technol.* **92** pp 65-69

[14] ZuliahaniA,RoziaizanN N, RozyantyAR, MohamadA F and NawawiW I 2016 Isolation and characterization of microcrystalline cellulose (MCC) from rice husk (RH) *MATEC Web Conferences* **47** 05013

[15] MacujaJ C O, RuedasL N andEspanaR C N 2015 Utilization of Cellulose from luffa cylindica fiber as binder in acetaminophen tablets *Adv. Environ. Chem.* **2015** pp 1-8

[16] JoharN, AhmadI and DufresneA 2012 Extraction, preparation and characterization of cellulose fibres and nanocrystals from rice husk *Ind. Crops and Prod.* **37** pp 93-99

[17] ChandrasekharS, SatyanarayanaK G, PramadaP N, RaghavanP and GuptaT N 2003 *J. Mater. Sci.* **38** pp 3159-3186

[18] HaafizM K M, EichhornS J, HassanA and JawaidM 2013 Isolation and characterization of microcrystalline cellulose from oil palm biomass residue *Carbohydr. Polym.* **93** pp 628-634

[19] AlemdarA and SainM 2008 Biocomposite from wheat straw nanofibers: morphology, thermal and mechanical properties *Compos. Sci. and Technol.* **68** pp 557-565

[20] NuruddinM, Chowdhury AHaque S A, RahmanM, FarhadS F and JahanM S 2011 *Cell. Chem. and Technol.* **45** pp 347-354

[21] FahmaF, IwamotoS, HoriN, IwataT and TakemuraA 2010 Isolation, perparation and characterization of nanofibers from (OPEFB) *Cellulose* **17** pp 977-985

IOP Conf. Series: Materials Science and Engineering **204** (2017) 012026 doi:10.1088/1757-899X/204/1/012026

Effect of Copper and Silicon on Al-5%Zn Alloy as a Candidate Low Voltage Sacrificial Anode

YudhaPratesa[1], DeniFerdian[1] and Inez Togina[1]

[1]Universitas Indonesia, Department of Metallurgy and Materials Engineering, Depok 16424

E-mail: yudha.pratesa@gmail.com, Deni@metal.ui.ac.id,ineztogina@gmail.com

Abstract. One common method used for corrosion protection is a sacrificial anode. Sacrificial anodes that usually employed in the marine environment are an aluminum alloy sacrificial anode, especially Al-Zn-In. However, the electronegativity of these alloys can cause corrosion overprotection and stress cracking (SCC) on a high-strength steel. Therefore, there is a development of the sacrificial anode aluminum low voltage to reduce the risk of overprotection. The addition of alloying elements such as Cu, Si, and Ge will minimize the possibility of overprotection. This study was conducted to analyze the effect of silicon and copper addition in Al-5Zn. The experiment started from casting the sacrificial anode aluminum uses electrical resistance furnace in a graphite crucible in 800°C. The results alloy was analyzed using Optical emission spectroscopy (OES), Differential scanning calorimetry, electrochemical impedance spectroscopy, and metallography. Aluminum alloy with the addition of a copper alloy is the most suitable and efficient to serve as a low-voltage sacrificial anode aluminum. Charge transfer resistivity of copper is smaller than silicon which indicates that the charge transfer between the metal and the electrolyte is easier t to occur. Also, the current potential values in coupling with steel are also in the criteria range of low-voltage aluminum sacrificial anodes.

1. Introduction

Oil and gas offshore exploration faced corrosion problems that caused by aggressive environment because it deals directly with sea water. Therefore, the structures for offshore, subsea pipelines and constructions in direct contact with sea water should be protected. Several ways can be used to reduce the corrosion rate. One common method used is the method of cathodic protection. Sacrificial anode applied in the marine environment is al-Zn-In alloy. Aluminum is an excellent sacrificial anode which used in seawater environments. However, passivation behavior of aluminum after contacted with its environment resulting in lower the output current. Therefore, aluminum is usually combined with other elements are like tin, indium, mercury, as an activator.

The conventional aluminum alloy sacrificial anode is Al-Zn-In, but electronegativity of these alloys can cause overprotection steel.[1,2] Therefore, development of low voltage sacrificial anode aluminum to reduce the risk of overprotection is urgently required nowadays. Therefore, there is a development of the low voltage sacrificial anode aluminum to lessen the possibility of overprotection [3,4].

The most common alloying elements added to the aluminum sacrificial anode is Zn. Zn added ranged between 2.5% -5.75%. The addition of Zn element is intended to damage the passive layer of aluminum [5]. Formation of second phase particles will trigger de-passivation of aluminum or damage

Content from this work may be used under the terms of the Creative Commons Attribution 3.0 licence. Any further distribution of this work must maintain attribution to the author(s) and the title of the work, journal citation and DOI.
Published under licence by IOP Publishing Ltd

to the passive layer of aluminum. It formed particles β-phase that caused damage to the passive layer of Al2O3 and increased efficiency of the sacrificial anode.
The addition of Cu will affect the strength and hardness of aluminum alloy. A Higher number of Cu, the corrosion resistance of aluminum will decrease. The main cause Cu may reduce the corrosion resistance of aluminum is not as solid-solution or a second phase formation, it resulting from a thin layer of Cu deposited on the surface of the alloy and formed a metallic copper cathode [6].
On the other hand, Silicon is a common alloy that is usually added to aluminum. Silicon is added to an alloy in aluminum sacrificial anodes are to reduce variation in corrosion and electrochemical properties[7]. The addition of silicon will improve the uniformity of the corrosion on the sacrificial anode because equiaxed fine grain and grain boundaries in which precipitates-silicon precipitates evenly distributed.

2. Methodology

Casting process of aluminum anode graphite crucible which under a heating temperature of 750°C and was held for 90 minutes. Remnants of impurities on the surface of the melt are removed before pouring process. Minimum purity each of element that used in this research was 99%. Each element is weighed to achieve a target composition of Al-5% Zn, Al-5Zn-(0.5-1%) Cu and Al-5Zn-(0.5-1%) Si. At least 2 Kg of sampling was achieved from each of casting process. The alloying process was done with holding the temperature at 750°C for 15 minutes. The stirring process was conducted for 20s to get a better homogenization. The casting product was characterized by several tests such as metallographic, optical emission spectroscopy and cyclic polarization test (Autolab PGSTAT302N). Cyclic polarization was done under 3.5%NaCl solution with scan rate 0.01mV/s, and potential range 1.0V to -1.0V.

3. Result

3.1 Optical Emission Spectroscopy

The chemical composition of each alloy was tested using Optical Emission Spectroscopy (OES). The result showed target composition has successfully achieved. Table 1 illustrates the result of OES testing.

Tabel 1 Chemical composition of casting product

Sample	Composition (wt.%)							
	Al	Zn	Cu	Fe	Si	Mn	Mg	In
Al-5Zn	94.9	4.90	<0.001	0.041	0.115	<0.001	<0.0001	<0.10
Al-5Zn-0.5Cu	93.4	5.16	0.571	0.058	<0.800	<0.0005	0.006	<0.10
Al-5Zn-1Cu	93.7	4.82	1.28	0.054	0.048	<0.001	0.0005	<0.01
Al-5Zn-0.5Si	94.8	4.55	0.561	0.034	<0.001	<0.001	<0.0001	<0.01
Al-5Zn-1Si	93.6	4.85	1.27	0.070	0.0037	0.0051	0.0031	<0.01

3.2 Metallographic

The metallographic test was conducted to compare the condition of metals before immersion and after immersion test. Based on the immersion test result, it can be seen that the material Al-5Zn-1Cu suffered severe corrosion than the Al-5Zn-1Si. Corrosion spread evenly at the entire alloy. It is because precipitate Cu at the secondary dendrite arm spacing region. Figure 1 showed corrosion of these alloy started from the dendrite secondaryarmspacing (SDAS) area. Precipitate in SDAS act as a cathode while area around the precipitates acted as anodic area.

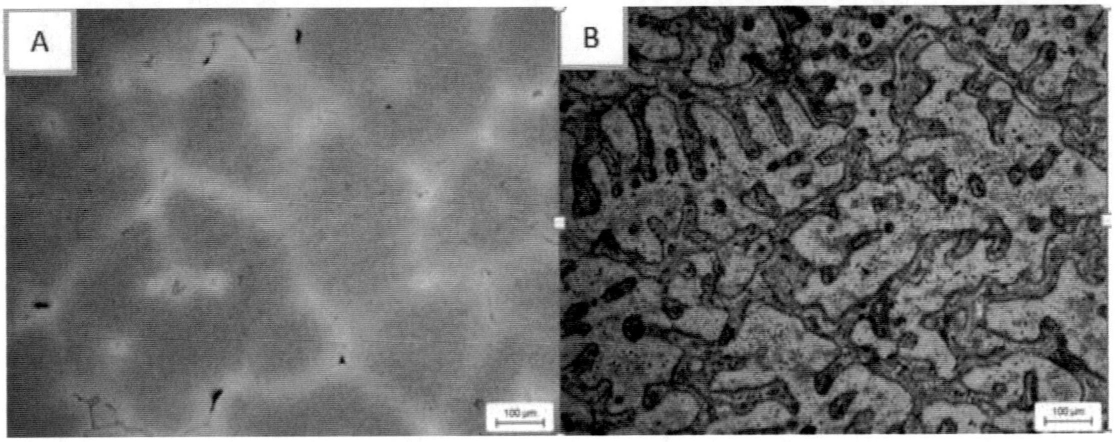

Figure 1 Condition of Al-Zn-Cu alloy before immersion (A) and after immersion for five days (B)

At the Al- 5Zn-1Si alloy, corrosion occurs around the intermetallic phase. However, corrosion attack is more benign compared with Al-5Zn-1Cu alloy. This condition showed intermetallic did not give sufficient effect compared with precipitate. Both of the alloys showed corrosion attack selectively at secondary dendrite arm spacing while the dendrite arm spacing are not attacked.

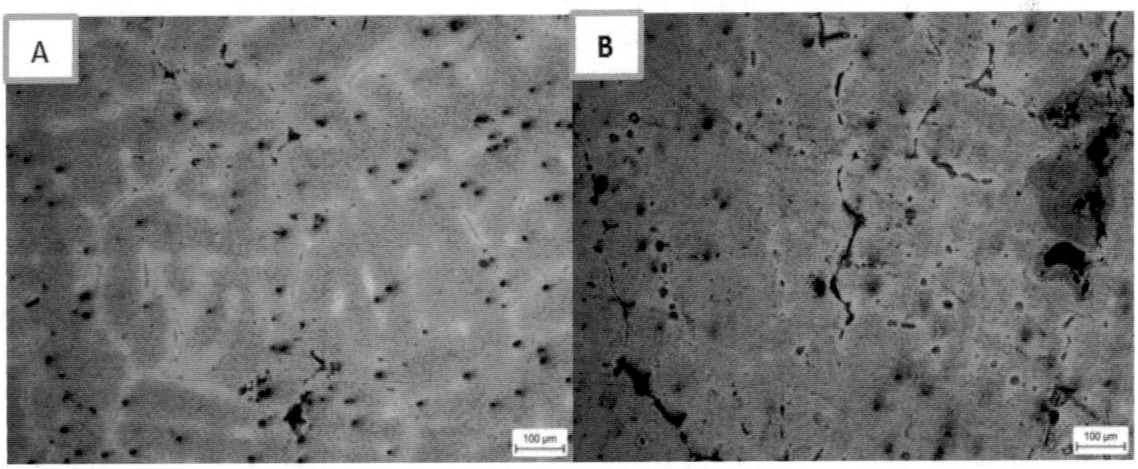

Figure 2 Condition of Al-Zn-Si alloy before immersion (A) and after immersion for five days (B)

3.3 Cyclic polarization

Based on cyclic polarization curve, materials with lower corrosion potential will be more susceptible to corrosion. Based on Figure 3 Al-5Zn alloy is the most vulnerable to corrosion due to the lowest potential than the other. On the other hand, Al-5Zn-1Cu has the highest potential than the other alloy candidate.

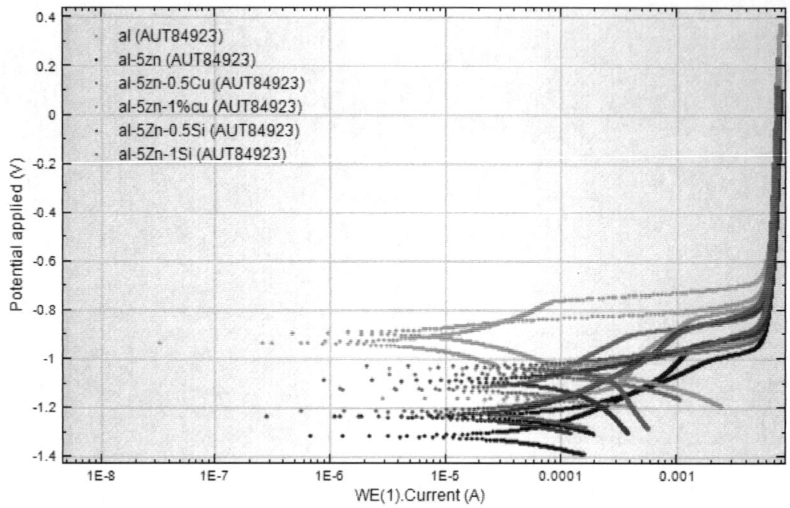

Figure 3 Cyclic polarization showed addition of copper would increase Ept umber

It can be seen quantitatively by Table 2. It can be seen that the pitting potential of Al-5Zn is equal to -0.94V vs. SCE and the most positive is the alloy Al-5Zn-1Cu -0.83 V vs. SCE. Even though Al-Zn1Cu increase the pitting potential of the alloy, it has the largest loop area. Higher loop area indicated corrosion would be propagated easier in the alloy due to longer time for oxide layer to re-passivated.

Tabel 2 Cyclic polarization analysis

Material Paduan	Eb (V vs SCE)	Ep (V vs SCE)	Eb-Ep (V vs SCE)
Al	-0.76	-0.83	0.07
Al-5Zn	-0.94	-1.14	0.2
Al-5Zn-0.5Cu	-0.89	-1.03	0.14
Al-5Zn-1Cu	-0.83	-0.99	0.16
Al-5Zn-0.5Si	-0.880	-0.99	0.11
Al-5Zn-1Si	-0.872	-0.97	0.10

4. Summary

Al-Zn-Cu alloy is the most suitable and efficient to be served as low-voltage sacrificial anode aluminium. It has higher corrosion rate and higher loop area. Precipitates Al-Zn-Cu give higher effect on corrosion than precipitate Al-Fe-Si

References

[1] Esaklul, K. A. & Ahmed, T. M. 2009 *J. Eng. Fail. Anal.* **16**, pp 1195–1202.
[2] Gurrappa, I. 2005.*J. Mat.Proc. Tech.* **166 (2)**, pp 256-267
[3] Lemieux E.J, Hogan E.A, Lucas K.E., Grolleau A.M. 2002,*Corrosion* 2002.
[4] Ma L., Qu B.W, Yan Y.G, Li K.Q 2014, Dev. Appl. Mater. 3.
[5] Muazu A, Yaro S.A., 2011, *J. Miner. Mater. Charact. Eng***10**, pp 185-198
[6] ASM Metals Handbook ,Volume 13: *Corrosion*. (1987).
[7] Wen J, He J, J. Wen, J. He, Lu X, 2011, *Corros. Sci.* **53**, pp 3861-3865